**Electronic
systems &
techniques**

Electronic systems & techniques

Second edition

K.F. Ibrahim

Longman
Scientific &
Technical

Longman Scientific & Technical,
Longman Group UK Limited,
Longman House, Burnt Mill, Harlow,
Essex, CM20 2JE, England
and Associated Companies throughout the world.

*Copublished in the United States with
John Wiley & Sons, Inc., 605 Third Avenue, New York, NY 10158*

© Longman Group UK Limited 1987, 1994

First published 1987
Second edition 1994

British Library Cataloguing in Publication Data
A CIP record for this book is available from the British Library

ISBN 0 582 21427 0

Library of Congress Cataloging-in-Publication Data
A catalog entry for this title is available from the Library of Congress

ISBN 0 470 23371 0 (USA only)

Set by 4 in Compugraphic Times 10/12
Printed in Malaysia by VVP

To TONIA, DINA and MARK

Contents

Preface   vii

Acknowledgement   ix

1   Direct current   1

2   Alternating current   9

3   Waveforms   13

4   Capacitive and inductive networks   19

5   Amplifiers and oscillators   31

6   Resonance   39

7   Transformers   42

8   Matching   49

9   Wave shaping   52

10   Power supplies I   56

11   Digital gates and systems   59

12   Microcomputers and other systems   66

13   Communication   71

14   Amplitude modulation   77

15   Frequency modulation   84

16   Pulse modulation   89

17   The cathode ray oscilloscope   91

18   Television   96

19   Useful exercises I   106

20   The semiconductor diode   120

21   Transistors   124

22 The common emitter amplifier 128

23 Common emitter amplifier — a.c. operation 134

24 The common base and common collector amplifiers 143

25 Photoelectric devices 146

26 Field effect transistors 148

27 More solid state devices 154

28 $R-C$ filters, clippers, and clampers 161

29 Power supplies II 171

30 Amplifiers 184

31 The operational amplifier 197

32 Multivibrators and timers 204

33 Oscillators 211

34 Logic circuits 217

35 Digital display 225

36 Microprocessor applications 230

37 Measuring instruments 233

38 Logical fault finding 238

39 Useful exercises II 258

Appendix I   Resistors 273

Appendix II   Capacitors 277

Appendix III   List of SI units 278

Appendix IV   Components and symbols 279

Appendix V   Notes on calculations 282

Appendix VI   The decibel 285

Appendix VII   Answers to useful exercises 288

Index 290

Preface

The book may be divided into four parts. Chapters 1–4 provide an introduction to the basic principles of electricity. Chapters 5–18 cover the variety of electronic systems in block diagram (or building block) format starting with the single stage amplifier right across to computer systems. Here the essential knowledge of the relationship between the various blocks or units is established.

Chapters 20–36 look inside the 'building blocks' that make up a system enquiring into their constituent parts and operation. The characteristics of components and the variety of circuitry used in electronics are explained with special attention to modern digital techniques.

Chapters 37 and 38 deal with instruments, testing, and fault finding. Analogue as well as digital and microprocessor test devices are included. In fault diagnosis, I have employed the functional approach where the state (cut-off or saturation; OFF or ON) of the unit-under-test is examined as the first step to identifying the faulty component.

In this second edition I have included two sets of eighty multiple choice questions. The first set (Chapter 19, Useful exercises I) covers the topics in Chapters 1 to 18. The second set (Chapter 39, Useful exercises II) covers topics in Chapters 20 to 38. I have also included new topics that have become necessary knowledge for basic electronic servicing such as analogue and digital converters, digital television receivers, timers and microprocessor test instruments.

Although the treatment I have adopted is non-mathematical, the systems and circuits are explained in depth to give the reader a sound understanding in electronics. Wherever possible I have indicated typical values of components that may be found in practice.

The book deals, to one extent or another, with the various aspects of modern electronics. It adequately covers the syllabus of Part I and Part II (core subjects and analogue electronics) of the City and Guilds 224 Course (Electronic Servicing). It is suitable for BTEC students at levels I, II and III taking electronic engineering. It is also useful to practising engineers who would like to back their practical background with theoretical knowledge.

The emphasis on digital electronics is intentional for it is in this field that recent advances have occurred and future developments lie. It has been my intention throughout the book to meet the challenge of conveying complex electronic concepts and systems in a manner that may be understood by those who have not had the opportunity of a mathematical background, but who may nonetheless wish to acquire the necessary skills to service modern electronic equipment. To this end I hope I have succeeded.

K.F. Ibrahim

Acknowledgement

We are indebted to Thorn Consumer Electronics Ltd. for permission to reproduce copyright material.

1 Direct current

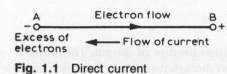

Fig. 1.1 Direct current

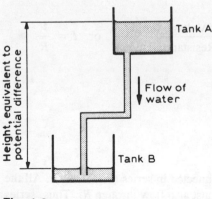

Fig. 1.2

Electric current

All matter consists of very small particles called atoms. These atoms are themselves composed of smaller particles known as **electrons** revolving around a central nucleus composed of one or more **protons** and **neutrons**. The electrons are negative particles while the protons are positive, neutrons are neutral.

If at a point A (Fig. 1.1) there is an excess of electrons as compared with point B, a potential difference (p.d.) or a voltage is said to exist between points A and B. If a conductor such as a wire is now used to connect A to B, the extra electrons at A will begin to flow towards B. This flow of electrons creates an electric current.

Point A with the extra electrons is described as having a negative potential and point B as having a positive potential.

The flow of electrons is similar to the flow of water from tank A to tank B shown in Fig. 1.2. The pipe is equivalent to the electrical conductor while the distance by which tank A is higher than tank B is equivalent to the potential difference between A and B.

Although electrons flow from the negative electrode towards the positive electrode, it is standard practice to say that current flows from positive to negative. This convention dates back to a time before current flow was fully understood. For the purposes of this book, conventional current flow (from positive to negative) will be used.

Potential difference or voltage is measured in volts. Current flow is measured in amperes.

Resistance

Just as a pipe opposes and restricts the flow of water through it, so an electrical resistance opposes and restricts the flow of current between two points. Resistance R is measured in ohms (symbol Ω).

Units

Although the basis units for current, voltage and resistance are the ampere, the volt and the ohm, other units are used which are a fraction or a multiple of the basic unit. These units are shown in Table 1.1.

1

Table 1.1

	Symbol	Unit
Current	I	ampere A
Voltage	V	volt V
Resistance	R	ohm Ω
milliamp	mA	$= \dfrac{1}{1000}$ A $= 10^{-3}$ A
microamp	μA	$= \dfrac{1}{1000}$ mA $= 10^{-3}$ mA or $\dfrac{1}{1\,000\,000}$ A $= 10^{-6}$ A
millivolt	mV	$= \dfrac{1}{1000}$ V $= 10^{-3}$ V
microvolt	μV	$= \dfrac{1}{1000}$ mV $= 10^{-3}$ mV or $\dfrac{1}{1\,000\,000}$ V $= 10^{-6}$ V
kilovolt	kV	$= 1000$ V $= 10^3$ V
kilo-ohm	kΩ	$= 1000\ \Omega$ $= 10^3$ V
mega-ohm	MΩ	$= 1000$ kΩ $= 10^3$ kΩ or $1\,000\,000\ \Omega$ $= 10^6\ \Omega$

Ohm's Law

A resistance by definition restricts the flow of current. The quantity of current that a resistor allows through depends on its resistance as well as the potential difference or voltage across it (Fig. 1.3). The greater the resistance, the smaller the current. On the other hand the higher the voltage, the greater the current. This relationship is known as Ohm's Law:

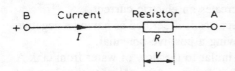

Fig. 1.3 Resistor in circuit

$$\text{Current (amperes)} = \frac{\text{Voltage (volts)}}{\text{Resistance (ohms)}} \quad \text{or} \quad I = \frac{V}{R}$$

Hence, by transposition,

$$R = \frac{V}{I} \quad \text{and} \quad V = IR$$

Resistors in series

R_1 and R_2 are two resistors connected in series (Fig. 1.4(a)). All the current that goes through R_1 must also flow through R_2. Thus, series resistors have a common current. Their voltages however are different.

EXAMPLE 1

If $R_1 = 2\ \Omega$, $R_2 = 6\ \Omega$, and $I = 3$ A, then

Voltage across R_1 is $V_1 = I \times R_1 = 3 \times 2 = 6$ V and
Voltage across R_2 is $V_2 = I \times R_2 = 3 \times 6 = 18$ V.

The total voltage V between A and B is equal to the sum of the voltages across R_1 and R_2.
Total voltage $V = V_1 + V_2 = 6$ V $+ 18$ V $= 24$ V.

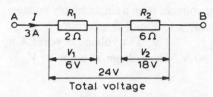

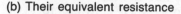

Fig. 1.4 (a) Two resistors in series

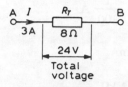

(b) Their equivalent resistance

Total resistance

R_1 and R_2 can be replaced by a single resistance which gives the same current for the same voltage between points A and B (Fig. 1.4(b)). Such an equivalent resistance is known as the total resistance R_T.

Total resistance $R_T = R_1 + R_2$.

Taking the resistor values of the example above,

$$R_T = R_1 + R_2 = 2 + 6 = 8\ \Omega$$

For a current of 3 A, the voltage can be calculated as

$$V = I \times R_T = 3 \times 8 = 24\ \text{V}.$$

This is the same total voltage as found by adding V_1 to V_2.

Three resistors in series

EXAMPLE 2

In Fig. 1.5 $R_1 = 1\ \text{k}\Omega$, $R_2 = 4\ \text{k}\Omega$, $R_3 = 10\ \text{k}\Omega$ and battery voltage $V = 15$ V. Calculate the current through the circuit and the voltage across each resistor.

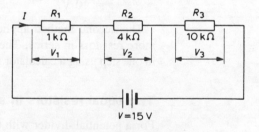

Fig. 1.5 Three resistors in series

Total resistance $R_T = R_1 + R_2 + R_3 = 15\ \text{k}\Omega$

Current $I = \dfrac{\text{Total voltage}}{\text{Total resistance}} = \dfrac{V}{R_T} = \dfrac{15\ \text{V}}{15\ \text{k}\Omega}$

$$= \frac{15\ \text{V}}{15 \times 1000\ \Omega} = \frac{1}{1000}\ \text{A} = 1\ \text{mA}$$

Voltage across R_1: $V_1 = IR_1 = 1\ \text{mA} \times 1\ \text{k}\Omega$

$$= \frac{1}{1000}\ \text{A} \times 1000\ \Omega = 1\ \text{V}$$

Voltage across R_2: $V_2 = IR_2 = 1\ \text{mA} \times 4\ \text{k}\Omega = 4\ \text{V}$
Voltage across R_3: $V_3 = IR_3 = 1\ \text{mA} \times 10\ \text{k}\Omega = 10\ \text{V}$.

The potential divider

As the example above shows, when two or more resistors are connected in series across a d.c. supply, different voltages appear across each resistor. Such an arrangement is called a potential divider

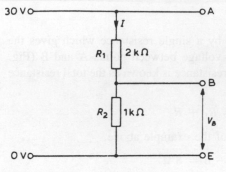

Fig. 1.6 The potential divider

and it is used to provide different voltages from a single supply voltage. A simple potential divider is shown in Fig. 1.6 where $R_1 = 2$ kΩ, $R_2 = 1$ kΩ and supply voltage $V = 30$ V. The voltage at point A is the same as the total voltage, 30 V. The voltage at point B, $V_B =$ voltage across R_2.

$$\text{Current through the circuit } I = \frac{\text{Total voltage}}{\text{Total resistance}}$$

$$\frac{30 \text{ V}}{1 \text{ k}\Omega + 2 \text{ k}\Omega} = \frac{30 \text{ V}}{3 \text{ k}\Omega} = \frac{30 \text{ V}}{3000 \text{ }\Omega} = \frac{1}{100} \text{ A}$$

$$= \frac{1}{100} \times 1000 = 10 \text{ mA}$$

$$\text{Voltage across } R_2 = IR_2 = 10 \text{ mA} \times 1 \text{ k}\Omega$$
$$= (10 \times 10^{-3}) \times (1 \times 10^3) = 10 \text{ V}.$$

Another way of calculating the voltage at point B is as follows:

$$V_B = \text{voltage across } R_2 = \frac{\text{Total voltage}}{\text{Total resistance}} \times R_2$$

$$= \frac{V}{R_1 + R_2} \times R_2 = \frac{30 \text{ V}}{2 \text{ k}\Omega + 1 \text{ k}\Omega} \times 1 \text{ k}\Omega = \frac{30}{3} \times 1$$

$$= 10 \text{ V}.$$

This second method applies to any potential divider with two or more resistors in series. The voltage at any point can be calculated in one step using a calculator without the need to find the current first.

Two equal resistors in series

For a potential divider with two equal resistors, the voltage divides equally across them.

Three equal resistors in series

EXAMPLE 3

The potential divider shown in Fig. 1.7 consists of three equal resistors of 1 kΩ each. Calculate the voltage at points A and B relative to E.

$$\text{Total resistance } R_T = R_1 + R_2 + R_3$$
$$= 1 \text{ k}\Omega + 1 \text{ k}\Omega + 1 \text{ k}\Omega = 3 \text{ k}\Omega$$

$$V_{AE} = \text{voltage across } R_1 = \frac{\text{Total voltage}}{\text{Total resistance}} \times R_1$$

$$= \frac{30 \text{ V}}{3 \text{ k}\Omega} \times 1 \text{ k}\Omega = 10 \text{ V}$$

$V_{BE} = $ voltage between points B and E

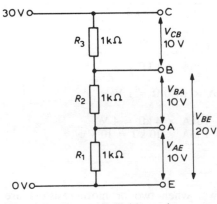

Fig. 1.7 Potential divider using three equal resistors

$$= \text{voltage across } (R_1 + R_2) = \frac{30 \text{ V}}{3 \text{ k}\Omega} \times (R_1 + R_2)$$

$$= \frac{30 \text{ V}}{3 \text{ k}\Omega} \times 2 \text{ k}\Omega = 20 \text{ V}.$$

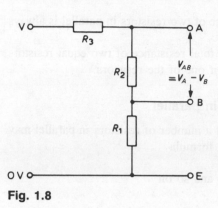

Fig. 1.8

Potential difference

The potential difference between any two points in a circuit is the difference in their respective voltages (relative to a common point (usually earth)). For instance, the p.d. between A and B in Fig. 1.8 is $V_{AB} = (V_A - V_B)$ where V_A is the voltage at point A and V_B is the voltage at point B. Both V_A and V_B are measured with respect to the zero line E. The voltage at any point in an electric circuit is usually measured with respect to the zero line, chassis or earth.

For example, if V_A is 5 V and V_B is 3 V, $V_{AB} = V_A - V_B = 5 - 3 = 2$ V (see Fig. 1.9(a)).

However, if positive and negative voltages are involved, then the p.d. is the sum of the two voltages. For example, if V_C is 3 V and V_D is -2 V, then $V_{CD} = V_C + V_D = 3 + 2 = 5$ V (Fig. 1.9(b)).

Generally, if the two voltages have the same polarity or sign, the p.d. between them is found by subtracting one from the other. But if the two voltages have different signs, they should be added together.

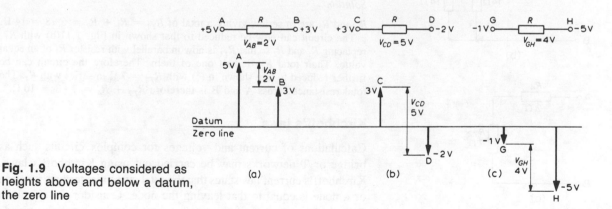

Fig. 1.9 Voltages considered as heights above and below a datum, the zero line

(a) (b) (c)

Resistors in parallel

Figure 1.10 shows two resistors, R_1 and R_2, connected in parallel. The current I taken from the battery is split at point A into I_1 through R_1 and I_2 through R_2. At B, I_1 and I_2 add up to form the total current I. Thus, total current $I = I_1 + I_2$.

On the other hand, the whole of the total voltage V is applied across each resistor. In other words,

Total voltage V = Voltage across R_1
= Voltage across R_2.

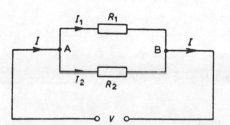

Fig. 1.10 Two resistors in parallel

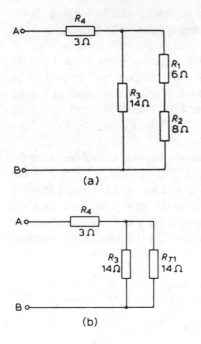

(a)

(b)

(c)

Fig. 1.11

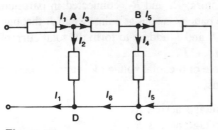

Fig. 1.12

Total resistance

The total or equivalent resistance R_T of two resistors in parallel is given by

$$R_T = \frac{R_1 \times R_2}{R_1 + R_2}$$

Note that the total resistance of two resistors in parallel is always **less** than the smaller resistor.

The case of $R_1 = R_2$. The total resistance of two equal resistors in parallel is half the value of one of the resistors.

Three or more resistors in parallel

In general, total resistance of a number of resistors in parallel may be found from the following formula

$$\frac{1}{R_T} = \frac{1}{R_1} + \frac{1}{R_2} + \frac{1}{R_3} \quad \text{and so on}$$

EXAMPLE 4

For the circuit shown in Fig. 1.11(a) find the total resistance.

Solution

R_1 and R_2 are in series, giving a total of $R_{T1} = R_1 + R_2 = 6+8 = 14 \; \Omega$.

The circuit can now be reduced to that shown in Fig. 1.11(b) with R_{T1} replacing R_1 and R_2 so that R_{T1} is now in parallel with resistor R_3 of an equal value. Their total R_{T2} is half one of them. Therefore the circuit can be further reduced to that shown in (c) with $R_{T2} = 7 \; \Omega$ in series with R_4. The total resistance across A and B is therefore $R_{T2} + R_4 = 7+3 = 10 \; \Omega$.

Kirchhoff's laws

Calculations of current and voltages for complex circuits such as bridge or T networks may be carried out using Kirchhoff's laws. Kirchhoff's current law states that the total current entering a junction or a node is equal to that leaving the node. Consider the circuit in Fig. 1.12 in which I_1 is the total current entering junction or node A while I_2 and I_3 are the currents leaving the node. It follows that

$$I_1 = I_2 + I_3$$

Similarly, for node B

$$I_3 = I_4 + I_5.$$

Suppose that $I_4 = 2$ mA and $I_5 = 3$ mA, then

$$I_3 = 2+3 = 5 \text{ mA}.$$

Assuming that $I_2 = 1$ mA, then

$$I_1 = I_2 + I_3 = 1+5 = 6 \text{ mA}.$$

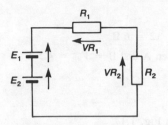

Fig. 1.13

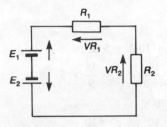

Fig. 1.14

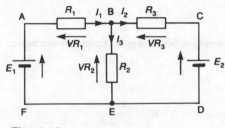

Fig. 1.15

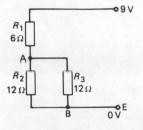

Fig. 1.16

Along the return path, at node C,

$$I_6 = I_4 + I_5 = 2+3 = 5 \text{ mA}.$$

And similarly, for node D,

$$I_1 = I_2 + I_6 = 1+5 = 6 \text{ mA}.$$

Kirchhoff's voltage law states that the total e.m.f. acting around a closed loop in a network is equal to the total voltage drops across all the resistors in the loop. For example, in the single-loop circuit in Fig. 1.13, the total e.m.f.s within the loop is $E_1 + E_2$, which is equal to the voltage drops across the two resistors, i.e.

$$E_1 + E_2 = VR_1 + VR_2.$$

If one of the e.m.f.s was reversed (Fig. 1.14), then E_2 is in the same direction (anti-clockwise) as VR_1 and VR_2, giving

$$E_1 - E_2 = VR_1 + VR_2 \text{ or}$$
$$E_1 = E_2 + VR_1 + VR_2.$$

For the multi-loop circuit in Fig. 1.15, taking loop ABEF we get:

$$E_1 = VR_1 + VR_2;$$

and loop ACDF gives

$$E_1 - E_2 = VR_1 + VR_3.$$

Going round loop BCDE we find that e.m.f. E_2 is in the same direction (clockwise) as VR_3, giving

$$E_2 + VR_2 = VR_3.$$

Power

When current passes through a resistance, energy is lost in the resistor in the form of heat. This is how the electric heater works, by converting electric energy into heat energy.

The unit for electric power is the watt, W. Power P is given by

$$P = V \times I$$

where V is the voltage in volts and I is the current in amps.

Using Ohm's law and substituting for V or I in terms of resistor R, the formula for power can be written as

$$P = I^2 \times R \quad \text{or} \quad P = \frac{V^2}{R} \text{ where } R \text{ is in ohms}$$

Worked examples

EXAMPLE 5

Calculate the voltage across R_3 in Fig. 1.16.

Solution

Total resistance between A and B = $\frac{1}{2} \times 12 = 6\ \Omega$.

Voltage across R_3 = voltage between A and B = V_{AB} = 4.5 V.

EXAMPLE 6

Calculate the voltage between X and Y in Fig. 1.17.

Solution

$R_3 + R_4 = 2+8 = 10\ \Omega$.

Total resistance between A and Y = 10/2 = 5 Ω.

Hence, voltage at point A, $V_{AY} = \dfrac{18}{4+5} \times 5 = \dfrac{18}{9} \times 5$

$= 10\ \text{V}$

Voltage between X and Y, $V_{XY} = \dfrac{10}{2+8} \times 8 = \dfrac{10}{10} \times 8 = 8\ \text{V}$

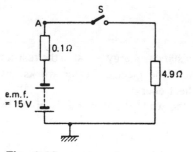

Fig. 1.17

EXAMPLE 7

A 2.5 Ω resistor is connected across a 50 V supply. Calculate the power dissipated.

Solution

The power dissipated in the resistor = $V^2/R = \dfrac{50 \times 50}{2.5}$

$= 1000\ \text{W} = 1\ \text{kW}$

EXAMPLE 8

Refer to Fig. 1.18. If the voltage reading at point A is 15 V when switch S is open, what is the reading if the switch is closed?

Solution

When S is closed, the total resistance of the circuit is 0.1 + 4.9 = 5 Ω.

Thus the current through the circuit is 15/5 = 3 A, giving a voltage across the 4.9 Ω load resistor of $3 \times 4.9 = 14.7$ V, which is the voltage at point A.

Fig. 1.18

2 Alternating current

A direct current flows in one direction only, but an alternating current (a.c.) is a current that continually changes direction.

A battery or a d.c. power supply provides direct current. A signal generator or the public mains supply delivers alternating current.

The sine wave

Alternating current can have many different shapes. These shapes are called waveforms. The most common waveform is the sine wave shown in Fig. 2.1. The sine wave has two equal peaks, a positive peak and a negative peak. The peak value is also known as the amplitude or the maximum value of the sine wave.

The peak-to-peak value of a sine wave is the potential difference between positive peak and the negative peak.

$$\text{Peak-to-peak value} = \text{Positive peak} + \text{Negative peak}$$
$$= 2 \times \text{peak value}$$

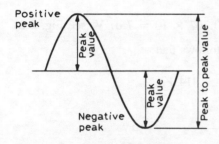

Fig. 2.1 Alternating current

Root mean square value

Unlike a direct current which has a constant value that can be used for all calculations, an alternating current varies with time and has no fixed value. To overcome this difficulty, a 'constant' value for the a.c. waveform called the root mean square value (r.m.s.) is used.

The r.m.s. value of an alternating current or voltage is the d.c. equivalent that provides the same power as the original a.c. waveform. Provided the r.m.s. value of an a.c. waveform is known, it can be used in all power calculations just as a d.c. voltage or current. For instance,

$$\text{d.c. power} = \text{d.c. current} \times \text{d.c. voltage}$$
$$\text{a.c. power} = \text{r.m.s. current} \times \text{r.m.s. voltage}$$

Root mean square values are always quoted for a.c. unless otherwise stated.

EXAMPLE 1

What is the resistance of a 1 kW domestic heater?

Solution

Domestic heaters use the mains voltage, 240 V r.m.s.

Power dissipated in the heater $= 1$ kW $= 1000$ W.

From $P = \dfrac{V^2}{R}$ it follows that $R = \dfrac{V^2}{P} = \dfrac{240 \times 240}{1000}$

$= 57.6\ \Omega$

Root mean square and peak value

The r.m.s. value of an a.c. waveform depends on the shape of the waveform.

For a sine wave, the r.m.s. value $= 0.707 \times$ the peak value.

This is true *only* for a sine wave.

For a sine wave with a peak voltage V_p of 10 V, its r.m.s. voltage is given by

$$V_{rms} = 0.707 \times V_p = 0.707 \times 10 = 7.07 \text{ V (see Fig. 2.2)}$$

From $V_{rms} = 0.707 \times V_p$ it follows that

$$V_p = \frac{1}{0.707} \times V_{rms} = 1.414\ V_{rms}$$

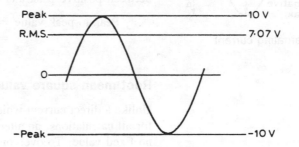

Fig. 2.2 The r.m.s. value of a sine wave

Direct current level of a.c. waveforms

So far we have considered a.c. waveforms that have no d.c. level. The two sine waves shown in Fig. 2.3 have the same peak value; sine wave A has a zero d.c. level as it goes positive as much as it goes negative, while sine wave B has a d.c. level of 5 V.

The d.c. level of an a.c. waveform is also referred to as the *average* or the *mean value*.

To find the d.c. level of a rectangular wave. Refer to Fig. 2.4.

1. First establish the zero line.
2. Find the area A_1 above the zero line for one cycle:
   $A_1 = 4 \times 1 = 4$.

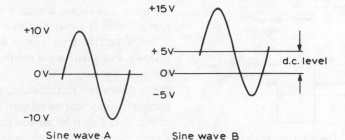

Fig. 2.3 The d.c. level of a.c.

Sine wave A   Sine wave B

Fig. 2.4

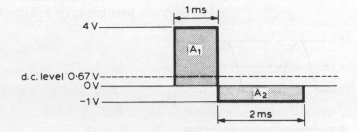

3. Find the area A_2 below the zero line for one cycle: $A_2 = 1 \times 2 = 2$.
4. Find the resultant area: $A_1 - A_2 = 4 - 2 = 2$.
5. Average voltage $= \dfrac{\text{Resultant area}}{\text{Periodic time}} = \dfrac{2}{3} = 0.67$ V.

Root mean square for complex waveforms

As stated earlier the formula r.m.s. $= 0.707 \times$ peak is true only for a sinusoidal waveform. Root mean square values for other waveforms may be calculated as follows:

1. Square the waveform over one cycle. Note that squaring a negative value turns it into a positive value.
2. Take the mean of the squared waveform.
3. Take the square root of the mean of the squared waveform, which gives the root mean square.

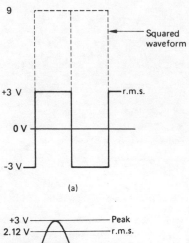

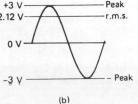

Fig. 2.5 Comparison between the r.m.s. values of a square wave and a sine wave

Hence, using the **square** wave shown in Fig. 2.5(a), squaring the positive half cycle gives $3 \times 3 = 9$. Squaring the negative half cycle gives $(-3) \times (-3) = 9$. The average over the squared cycle is therefore 9. Hence the r.m.s. voltage is $\sqrt{9} = 3$ V.

Comparing this with a sine wave having positive and negative peaks of $+3$ V and -3 V respectively, with an r.m.s. of $0.707 \times 3 = 2.12$ V (Fig. 2.5(b)), we find that the square wave has a higher r.m.s. value than the sine wave. This is because the area under the square waveform is larger than that under the sinusoidal waveform when both are alternating between the same positive and negative values. Note: for a square wave r.m.s. value = peak value.

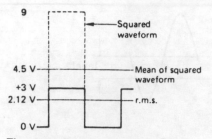

Fig. 2.6 Root mean square for positive-going square wave

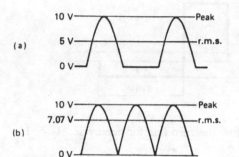

Fig. 2.7 (a) The r.m.s. value of a half-wave rectified sine wave is $\frac{1}{2} \times$ peak (b) The r.m.s. value of a full-wave rectified sine wave is $0.7 \times$ peak

Figure 2.6 shows a square wave which is wholly positive. The r.m.s. value is smaller in this case.

For a **rectified** waveform the r.m.s. value may be deduced as follows. For a half-wave rectified sine wave, the r.m.s. value is half the peak value.

For full-wave rectification, the r.m.s. is the same as that for a complete sinusoidal waveform, namely $0.707 \times$ peak value as shown in Fig. 2.7. This is because as far as r.m.s. calculations are concerned, a negative-going half cycle is identical to a positive-going half cycle.

Note that the d.c. or mean value of a waveform is merely the average value over one cycle and thus bears no relation to the r.m.s. value.

3 Waveforms

The variation of a voltage or a current with time can be represented by a waveform.

Direct current being steady, constant current has a straight line waveform (Fig. 3.1(a)). Alternating current on the other hand can have numerous shapes. The a.c. waveform is one which alternates between a maximum value and a minimum value and back to a maximum and so on. They are thus repetitive waveforms (Fig. 3.1(b)). Some such waveforms are shown in Fig. 3.2.

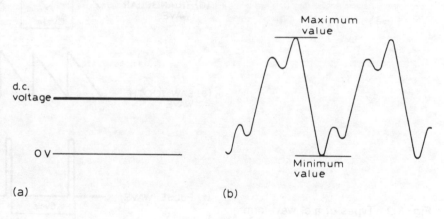

(a)   (b)

Fig. 3.1 (a) d.c. waveform (b) a.c. waveform

A cycle

The part of the a.c. waveform that repeats itself is known as the cycle of the waveform. Referring to Fig. 3.2(a), if we take point A as the beginning of a cycle, then point B will be the end of that cycle and the beginning of the next. The part of the waveform from A to B repeats itself and thus is one cycle.

Frequency

The number of times a cycle repeats itself in one second is known as the frequency of the waveform. The unit for frequency is the hertz (Hz).

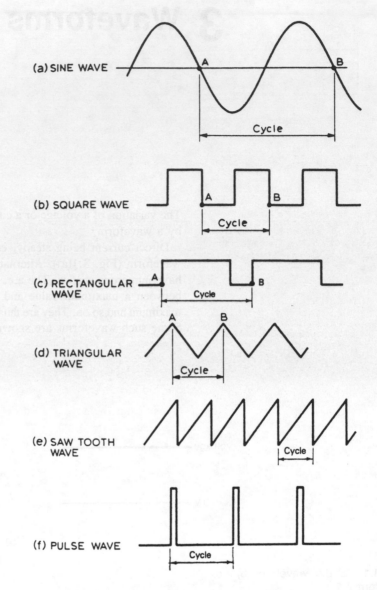

(a) SINE WAVE

(b) SQUARE WAVE

(c) RECTANGULAR WAVE

(d) TRIANGULAR WAVE

(e) SAW TOOTH WAVE

(f) PULSE WAVE

Fig. 3.2 Types of a.c. waveform

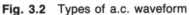
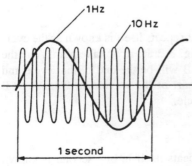

Fig. 3.3 Frequency

For example, if a waveform repeats itself once every second, its frequency is 1 Hz. If it repeats itself 10 times every second (Fig. 3.3), its frequency is then 10 Hz.

Periodic time

The time taken to complete one cycle is known as the periodic time, or period, of the waveform. For instance, a cycle which takes one second to complete has a periodic time of 1 s. If it takes half a second to complete, its periodic time is 0.5 s.

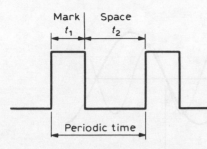

Mark | Space
t_1 | t_2

Periodic time

Fig. 3.4 Mark-to-Space ratio less than one

Mark and Space

One cycle of a rectangular wave can be divided into a Mark and Space as shown in Fig. 3.4. The ratio of Mark time to Space time is known as the Mark-to-Space ratio. If t_1 is the Mark time and t_2 is the Space time then,

$$\text{Mark-to-Space ratio} = \frac{\text{Mark time}}{\text{Space time}} = \frac{t_1}{t_2}$$

The periodic time of the waveform being the time for one complete cycle is

$$\text{Mark} + \text{Space} = t_1 + t_2$$

For a Mark-to-Space ratio of one,

$$\text{Mark time } t_1 = \text{Space time } t_2$$

It follows that

$$\text{Periodic time} = 2 \times \text{Space time} = 2 \times \text{Mark time}$$

The waveform is therefore a square wave.

Units: frequency f

hertz Hz

kilohertz kHz $= 1000 \text{ Hz} = 10^3 \text{ Hz}$

Megahertz MHz $= 1000 \text{ kHz} = 10^3 \text{ kHz} = 10^6 \text{ Hz}$

Units: periodic time T

second s

$$\text{millisecond ms} = \frac{1}{1000} \text{ s} = 10^{-3} \text{ s}$$

$$\text{microsecond } \mu\text{s} = \frac{1}{1000} \text{ ms} = 10^{-3} \text{ ms} = 10^{-6} \text{ s}$$

Frequency and period time

Taking the two waveforms A and B in Fig. 3.5 it can be seen that the frequency of B is double that of A. The periodic time of B on the other hand is half that of A. As the frequency of a waveform increases, so its periodic time decreases and vice versa.

$$\text{Frequency} = \frac{1}{\text{Periodic time}} \text{ Hz} \quad \text{or} \quad f = \frac{1}{T} \text{ Hz and}$$

$$\text{Periodic time} = \frac{1}{\text{Frequency}} \text{ s} \quad \text{or} \quad T = \frac{1}{f} \text{ s}$$

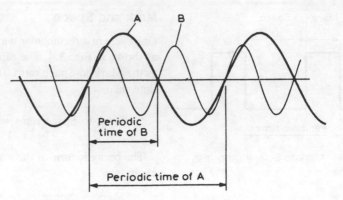

Fig. 3.5

The following table is a useful table of conversion and should be committed to memory.

frequency f	1 Hz	1 kHz	1 MHz
periodic time T	1 s	1 ms	1 μs

Sound waves

Sound waves are those produced in the air when a person is speaking, or from a loudspeaker, a tuning fork, or a pneumatic drill, etc. They are variations of air pressure and require the air as a medium to travel through.

The *volume* of a sound wave indicates its loudness. The *tone* indicates its frequency. By changing the frequency, the tone of the sound is varied.

Audio frequency

The range of sound frequency that can be heard by the human ear is known as the audio frequency (a.f.) range. The a.f. range extends from 20 Hz to 20 kHz. Frequencies below 20 Hz or above 20 kHz cannot be heard by the human ear. This is why the sound produced by special whistles used to call sheep dogs is inaudible by humans. The frequency of these sound waves is above 20 kHz to which the ear of a dog is able to respond.

Pure and instrumental tones

A pure tone is a single sine wave having a single frequency (Fig. 3.2(a)). An instrumental tone is a complex wave composed of a number of different sine waves having different frequencies added together (Fig. 3.1(b)). These are examples of the waveforms produced in speech or music.

Harmonics

By adding a number of different sine waves to one another, a complex wave is produced. Conversely, a complex wave can be found to contain a number of pure sine waves added together. These pure sine waves are known as the fundamental and the harmonics. Any complex waveform can be analysed into the following components:

1. *The fundamental*. A pure sine wave having the same periodic time as the original complex wave.
2. A number of *harmonics*. Pure sine waves having frequencies that are a multiple of the fundamental, e.g. if the fundamental frequency is 100 Hz, then

   the 2nd harmonic = $2 \times 100 = 200$ Hz
   the 3rd harmonic = $3 \times 100 = 300$ Hz
   the 4th harmonic = $4 \times 100 = 400$ Hz and so on.

Higher harmonics have smaller amplitude than the lower ones. For this reason the higher harmonics are usually neglected.

Pitch

The pitch of a sound wave indicates the part of the a.f. range within which its frequency falls. A high-pitched sound falls in the upper half of the a.f. scale while a low-pitched sound falls in the lower half. Women generally have higher pitched voices than men. A drum produces a sound of low pitch while a flute produces high-pitched notes. For a complex wave, the fundamental frequency determines its pitch.

Quality

The quality of sound is determined by the number of harmonics of the instrumental wave that are reproduced without any distortion.

Some complex waves

1. Fundamental + 3rd harmonic only (Fig. 3.6)
2. Fundamental + 2nd harmonic only (Fig. 3.7)

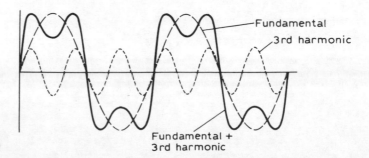

Fig. 3.6 Fundamental + 3rd harmonic (The waveform approximates that of a square wave)

Fundamental
3rd harmonic

Fundamental + 3rd harmonic

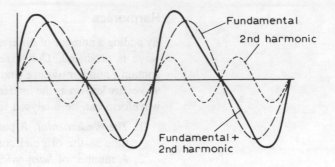

Fig. 3.7 Fundamental + 2nd harmonic (The waveform approximates that of a saw-tooth wave)

The harmonics of a square wave

A square wave consists of the fundamental together with an infinite number of **odd** harmonics only. For example, a square wave having a frequency of 1 kHz contains the following components:

fundamental   = 1 kHz
3rd harmonic = 3 × 1 = 3 kHz
5th harmonic = 5 × 1 = 5 kHz
7th harmonic = 7 × 1 = 7 kHz and so on.

It will be noticed that complex waves containing only odd harmonics have sharp rising and falling edges. The greater the number of these odd harmonics, the nearer we get to a square wave.

The harmonics of a saw-tooth wave

A saw-tooth wave consists of the fundamental together with an infinite number of **even** harmonics only. For example a saw-tooth wave having a frequency of 1 kHz has the following components:

fundamental   = 1 kHz
2nd harmonic = 2 × 1 = 2 kHz
4th harmonic = 4 × 1 = 4 kHz
6th harmonic = 6 × 1 = 6 kHz and so on.

4 Capacitive and inductive networks

The capacitor

A capacitor consists of two conductor plates facing each other with an insulating material called the **dielectric** separating them (Fig. 4.1). Common dielectric materials are air, paper, mica, etc.

Charging a capacitor

The basic property of a capacitor is its ability to store electric energy in the form of an electric charge.

Figure 4.2(a) shows a battery connected across a capacitor via a switch. When the switch is closed as at (b), the positive electrode of the battery draws electrons from plate A, giving the plate a positive charge. The negative electrode drives electrons to plate B, making it negative by an equal amount. This flow of electrons is known as

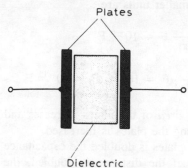

Plates

Dielectric

Fig. 4.1 The capacitor

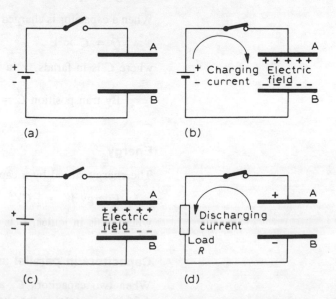

(a)

(b)

+ Charging Electric
- current   field

(c)

Electric
field

(d)

Discharging
current

Load
R

Fig. 4.2 Charging and discharging a capacitor

the charging current, which continues to flow until the voltage across the capacitor equals the e.m.f. of the battery. The capacitor is then said to be fully charged. The symbol for **charge** is Q and its unit is the Coulomb.

When a capacitor is charged, the potential difference created between the plates produces an electric field.

When the switch is opened as at (c), the capacitor sustains the charge within the dielectric which maintains the electric field between its two plates. The electrical field of a charged capacitor will collapse when the capacitor is discharged through a load resistance, as at (d).

Capacitance

The ability of a capacitor to store a charge is known as its capacitance C. The unit of capacitance is the farad F. The farad is a very large unit and therefore hardly used. Smaller units are

$$\text{microfarad } \mu F = \frac{1}{1\ 000\ 000} \text{ F} = 10^{-6} \text{ F}$$

$$\text{picofarad pF} = \frac{1}{1\ 000\ 000} \mu F = 10^{-6} \mu F = 10^{-12} \text{ F}$$

Capacitance C increases as the *area* of the plates increases and decreases as the *distance* separating the plates is increased.

For example, if the area of the plates is doubled the capacitance will double in value. However, if the distance is doubled, the capacitance is halved.

Charge, capacitance and voltage

When a capacitor is charged to a voltage V, the charge Q is given by

$$Q = C \times V$$

where C is in farads, V in volts, and Q in coulombs.

By transposition $C = \dfrac{Q}{V}$ and $V = \dfrac{Q}{C}$

Energy

The energy stored by a capacitor is given by

$$\text{Energy } W = \tfrac{1}{2} CV^2$$

where W is in joules, C is in farads, and V is in volts.

Capacitors in parallel and series

When two capacitors C_1 and C_2 are connected in parallel (Fig. 4.3(a)) their total is the sum of the two, i.e.

Total capacitance $C_T = C_1 + C_2$

When connected in series (Fig. 4.3(b)) the total is smaller than either one:

Total capacitance $C_T = \dfrac{C_1 C_2}{C_1 + C_2}$

For example, if $C_1 = C_2$, their total when connected in series is half one of them; $\frac{1}{2}C_1 = \frac{1}{2}C_2$.

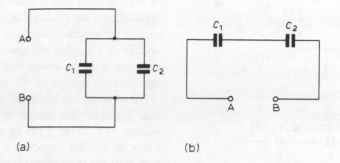

Fig. 4.3 (a) Capacitors in parallel (b) Capacitors in series

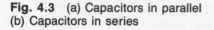

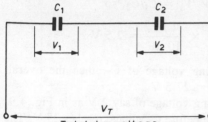

Fig. 4.4 Voltage across series capacitors

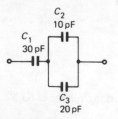

Fig. 4.5

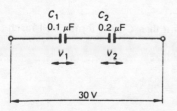

Fig. 4.6

Voltage across series capacitors

In the circuit shown in Fig. 4.4, C_1 is in series with C_2 with a d.c. voltage V_T connected across them. The total voltage V_T will divide between C_1 and C_2 so that the smaller capacitor will sustain a higher voltage across it and vice versa. The sum of V_1 (across C_1) and V_2 (across C_2) is always equal to the total voltage V_T.

Total voltage $V_T = V_1 + V_2$ with $\dfrac{V_1}{V_2} = \dfrac{C_2}{C_1}$

In general, when a number of capacitors are connected in series across a d.c. supply, the voltage across each capacitor is inversely proportional to its capacitance. For two capacitors in series (Fig. 4.4) the voltages across C_1 and C_2 are given as

$$V_1 = \frac{V_T}{C_1 + C_2} \times C_2; \quad V_2 = \frac{V_T}{C_1 + C_2} \times C_1$$

EXAMPLE 1

In Fig. 4.5 the total capacitance is calculated as follows. Total capacitance of parallel combination is

$C_2 + C_3 = 10 + 20 = 30$ pF

Since C_1 is also 30 pF, then the total capacitance is $\frac{1}{2} \times 30 = 15$ pF.

EXAMPLE 2

In Fig. 4.6 the voltage across capacitor C_1 is

$$\frac{V_T}{C_1+C_2} \times C_2 = \frac{30}{0.1+0.2} \times 0.2 = 20 \text{ V}$$

giving a voltage across $C_2 = 30 - 20 = 10$ V.

Working voltage

Every capacitor has a maximum voltage it can sustain before the dielectric breaks down. This voltage is known as the working voltage of the capacitor and it must never be exceeded. When used in a.c. applications, the peak voltage applied across the capacitor must not exceed the specified working voltage of the capacitor. The working voltage of a number of capacitors connected in parallel is that of the capacitor with the lowest working voltage. For example in Fig. 4.7, the working voltage of the parallel circuit is 25 V.

For capacitors in series the working voltage is more difficult to work out. Consider the circuit in Fig. 4.8 where C_1 (1 μF, working voltage WV = 25 V) is in series with C_2 (10 μF, WV = 10 V). Since the smaller capacitor C_1 will sustain a higher voltage than C_2, then start the calculations with C_1 at its specified working voltage of 25 V. Thus $V_1 = 25$ V. But $V_2/V_1 = C_1/C_2$, so

$$V_2 = V_1 \times \frac{C_1}{C_2} = 25 \text{ V} \times \frac{1 \ \mu\text{F}}{10 \ \mu\text{F}} = 2.5 \text{ V}$$

Since V_2 is below the working voltage of C_2, then the overall working voltage is $25 + 2.5 = 27.5$ V.

However, if C_2 had a working voltage of say 2 V as in Fig. 4.9, it will reach its working voltage before C_1 voltage reaches 25 V. The overall working voltage is calculated as follows:

$$V_2 = 2 \text{ V and } V_1 = V_2 \times \frac{C_2}{C_1} = 2 \times \frac{10}{1} = 20 \text{ V}$$

giving an overall working voltage of $20 + 2 = 22$ V.

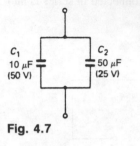

Fig. 4.7

C_1
10 μF
(50 V)

C_2
50 μF
(25 V)

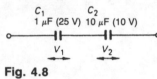

Fig. 4.8

C_1
1 μF (25 V)

C_2
10 μF (10 V)

V_1 V_2

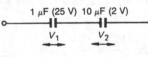

Fig. 4.9

1 μF (25 V) 10 μF (2 V)

V_1 V_2

EXAMPLE 3

Refer to Fig. 4.10 with C_1 and C_2 having the same working voltage of 60 V. What is the maximum voltage that may be applied?

Solution

Since C_1 sustains a higher voltage across it than C_2, the former will reach its working voltage first,

with $V_1 = 60$ V,

$$V_2 = \frac{C_1}{C_2} \times V_1 = \frac{2}{6} \times 60 = 20 \text{ V}$$

The maximum voltage that may be applied $= 60 + 20 = 80$ V.

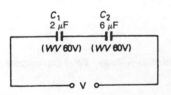

C_1
2 μF

C_2
6 μF

(WV 60V) (WV 60V)

 V

Fig. 4.10

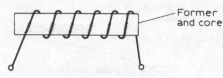

Fig. 4.11   Simple inductor

Fig. 4.12   The inductor

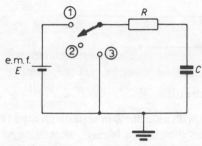

Fig. 4.13

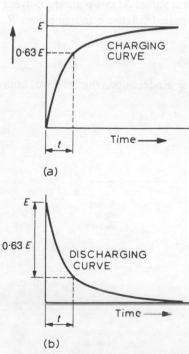

Fig. 4.14   Charging and discharging curves, where t is the time constant

The inductor

The inductor is simply a conductor wound as shown in Fig. 4.11. Its symbol is shown in Fig. 4.12. While a capacitor opposes a change of voltage across it, an inductor opposes a change of current through it. In other words if a step current is applied to a circuit with an inductor, current will slowly build up to its maximum.

This ability of an inductor to oppose a change in current is known as its inductance. The symbol of inductance is L and its unit is the henry H. In smaller units,

$$\text{milli-henry mH} = \frac{1}{1000}\,\text{H} = 10^{-3}\,\text{H}$$

$$\text{micro-henry } \mu\text{H} = \frac{1}{1000}\,\text{mH} = 10^{-3}\,\text{mH} = \frac{1}{1\,000\,000}\,\text{H}$$

$$= 10^{-6}\,\text{H}$$

Time constant for CR circuit

Figure 4.13 shows a capacitor in series with a resistor connected across a battery via a switch. When the switch is in position (1), the capacitor gradually charges up through resistor R until the voltage across it is equal to E, the e.m.f. or voltage of the battery.

The manner in which the capacitor charges up is shown in Fig. 4.14(a) along an exponential curve.

The time taken for the capacitor to charge to 0.63 of the maximum, in this case 0.63 E, is known as the time constant of the circuit.

Referring back to Fig. 4.13, if the switch is now moved to position (2), the capacitor will store the energy, keeping the voltage across it constant. When the switch is moved to position (3), the capacitor begins to discharge through resistor R to earth. The voltage across the capacitor gradually falls to zero. The manner in which the discharge takes place is shown in Fig. 4.14(b). In this case the time constant is the time it takes the voltage across the capacitor to fall by 0.63 of its maximum value.

For either charging or discharging a capacitor through a resistor R, the time constant is given by

$$t = C \times R$$

where t is the time constant in seconds, C is the capacitance in farads, R is the resistance in ohms.

For example if $C = 10\ \mu\text{F}$ and $R = 10\ \text{k}\Omega$ then the time constant of the circuit is

$$t = CR = 10\ \mu\text{F} \times 10\ \text{k}\Omega = 10 \times 10^{-6} \times 10 \times 10^{3}$$
$$= 10 \times 10 \times 10^{-3} = 10^{2} \times 10^{-3} = 10^{-1} = 0.1\ \text{s}.$$

Figure 4.15 shows charging curves for a short and a long time constant.

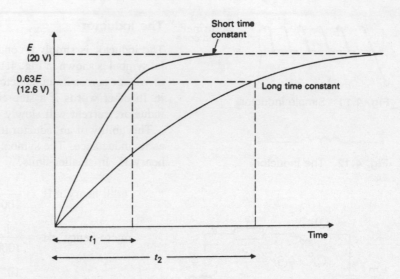

Fig. 4.15 Changing curves for a short and a long time constant

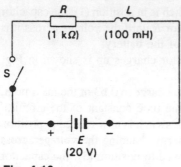

Fig. 4.16

Time constant for an LR circuit

Consider the circuit in Fig. 4.16 with inductor L in series with a 1 kΩ resistor R. When switch S closes the current is zero, attempting to increase sharply. However, since the inductor opposes any change of current through it, the current I through the circuit will increase exponentially, as shown in Fig. 4.17. The current will continue to increase until it reaches its maximum value. At this point the current stops increasing, with the whole applied voltage E falling across R. Steady-state current

$$= E/R = 20/1 \text{ k}\Omega = 20 \text{ mA}$$

The rate of change of current is dependent upon the values of both R and L.

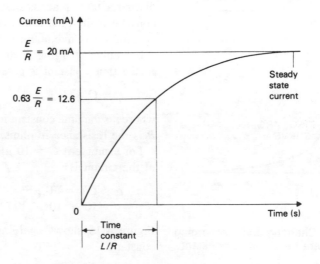

Fig. 4.17 Exponential increase of a current through an inductor

The time it takes the current to reach 0.63 of its maximum steady state is known as the time constant of the circuit. This may be calculated as L/R, where L is in henries and R is in ohms, and the time constant is in seconds. Using the values in the diagram,

$$\text{time constant} = \frac{L}{R} = \frac{100 \text{ mH}}{1 \text{ k}\Omega} = \frac{0.1 \text{ H}}{1000 \text{ }\Omega}$$

$$= 10^{-4} \text{ s} = 0.1 \text{ ms}$$

Note that for a large R, the time constant L/R is small, with the current changing at a faster rate.

Direct current resistance

An inductor presents no resistance to the flow of d.c. current other than the very small resistance of the wire from which it is made. Thus an inductor has zero or very small d.c. resistance and can be considered as a short circuit for d.c. A capacitor on the other hand, due to the insulating dielectric, presents an infinite or very large resistance to direct current and can be considered as an open circuit for d.c.

Phasor representation

A sine wave may be represented by a phasor OA rotating anti-clockwise at an angular velocity $\omega = 2\pi f$, where f is the frequency of the waveform (Fig. 4.18). As the phasor rotates, the height of its tip (point A) above the horizontal axis describes the sine wave shown. One complete revolution of the phasor (360° or 2π) represents one complete cycle. Thus half a revolution (180° of π) represents half a cycle, and so on. The time axis of the waveform may, therefore, be used to represent the angle through which the phasor has moved, as shown in the figure. The positive peak value is therefore at 90° ($\frac{1}{4}$ cycle) and the negative peak is at 270° ($\frac{3}{4}$ cycle).

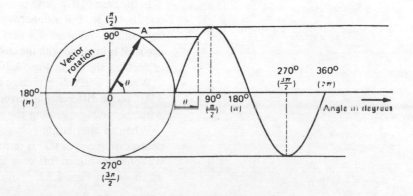

Fig. 4.18 Phasor representation of a sine wave

Now consider the two sine waves A and B, represented by phasors OA and OB respectively, in Fig. 4.19(a). If both sine waves have the same frequency, then both OA and OB will rotate at equal angular velocity $\omega = 2\pi f$. This means that the angular difference θ between OA and OB does not change. OA is said to be *leading* OB by an angle of θ, or OB is *lagging* OA by an angle of θ. The sine waves are as shown in Fig. 4.19(b).

If these two sine waves are added together, their resultant will be another sine wave having the same frequency f but a different amplitude. The resultant sine wave may be represented by a separate phasor OT which is the phasor sum of OA and OB as shown in Fig. 4.19(c). The resultant waveform leads OB by angle α and lags OA by angle γ.

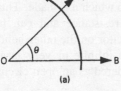

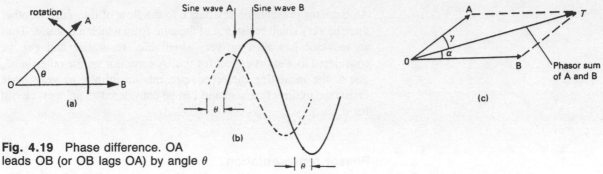

Fig. 4.19 Phase difference. OA leads OB (or OB lags OA) by angle θ

As we shall see in the remainder of this chapter, phasor representation is a very useful device in analysis and calculation in a.c. networks.

Phase difference

When a voltage V is applied across a resistor R a current I flows through it. The voltage and current are distinct electrical qualities with their own units and waveforms.

In the case of d.c. networks (Chapter 1) the question of waveforms does not arise. For a.c. networks the relationship between the current and voltage waveforms is very important. For a resistor R the voltage across it is in phase with the current through it, i.e. the phase difference is zero as shown in Fig. 4.20(b) and (c).

When an alternating voltage is applied across an inductor (Fig. 4.21) a 90° phase difference is introduced between the voltage and current, with the voltage *leading* the current.

When an alternating voltage is applied across a capacitor a similar phase difference of 90° is introduced between the voltage and current waveforms, but in this case the voltage will be *lagging* the current as shown in Fig. 4.22.

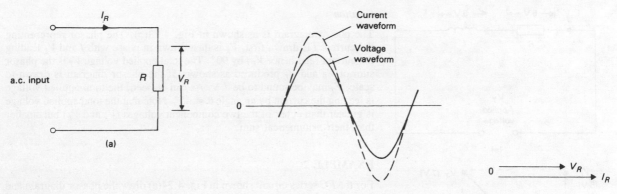

Fig. 4.20

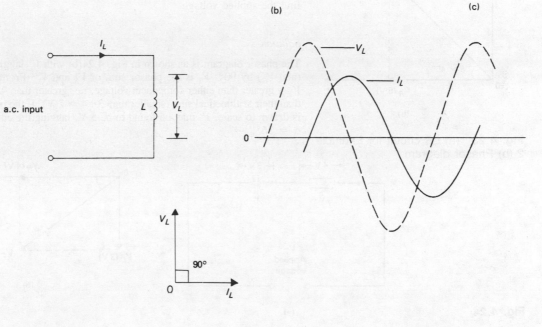

Fig. 4.21

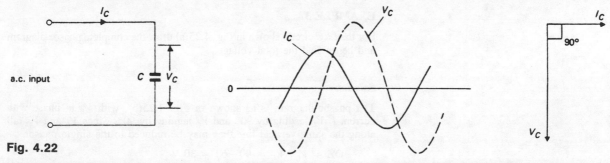

Fig. 4.22

EXAMPLE 1

For the *LR* circuit shown in Fig. 4.23(a) draw the phasor diagram, and hence find the total applied voltage V_T.

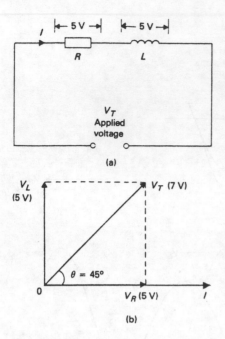

Fig. 4.23 (a) *LR* circuit for Example 2 (b) Phasor diagram

Solution

The phasor diagram is as shown in Fig. 4.23(b). The phasor representing the current I is drawn first. V_R is then drawn in phase with I and V_L leading the current (and hence V_R) by 90°. The total applied voltage V_T is the phasor sum of V_R and V_L produced as shown. If the phasor diagram is drawn to scale, V_T may be found to be 7 V. As can be seen, the total applied voltage is leading the current by an angle $\theta = 45°$. Note that the total applied voltage is greater than either of the two component voltages (V_R and V_L) but smaller than their arithmetical sum.

EXAMPLE 2

For the *RC* series circuit shown in Fig. 4.24(a) draw the phasor diagram and find the applied voltage.

Solution

The phasor diagram is as shown in Fig. 4.24(b) with V_C lagging the current (and V_R) by 90°. V_T is the phasor sum of V_R and V_C. From the diagram, V_T is greater than either component voltage, i.e. greater than 4 V but smaller than their arithmetical sum (smaller than $3+4 = 7$ V). If the phasor diagram is drawn to scale, V_T may be found to be 5 V, lagging the current by angle $\alpha = 36°$.

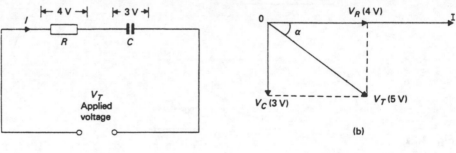

Fig. 4.24

(a)

EXAMPLE 3

For the *LRC* circuit shown in Fig. 4.25(a) draw the complete phasor diagram and hence find the total voltage.

Solution

The phasor diagram is as shown in Fig. 4.25(b), with V_R in phase with current I, V_L leading by 90° and V_C lagging by 90°. Since V_L and V_C fall along the same vertical line they may be reduced to the single phasor

$$OZ = V_L - V_C = 90 - 60 = 30 \text{ V}$$

as shown in Fig. 4.25(c). The total voltage V_T is, therefore, the phasor sum of V_R and OZ. $V_T = 50$ V leading the current by angle $\alpha = 36°$.

Reactance

Inductors and capacitors offer opposition to the flow of alternating

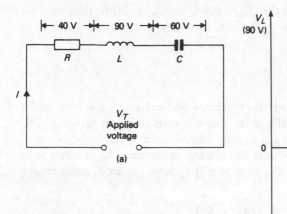

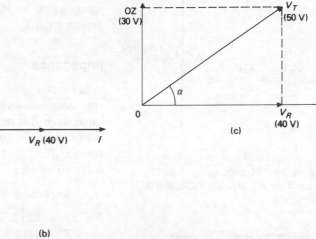

Fig. 4.25 (a) *LRC* circuit for Example 3 (b) Phasor diagram (c) Resultant phasor

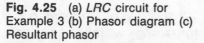

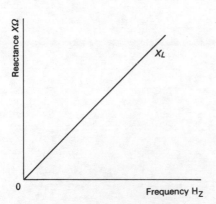

Fig. 4.26 Graph showing how the reactance of an inductor X_L changes with frequency

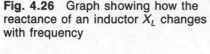

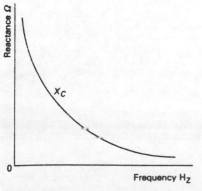

Fig. 4.27

current. This a.c. resistance is known as reactance X and is measured in ohms. Reactance depends on the value of the inductor or the capacitor as well as the frequency of the a.c. waveform.

An inductor has an inductive reactance X_L,

$$X_L = 2\pi f L$$

where f is the frequency in Hz and L is the inductance in henries.

Since $\omega = 2\pi f$, then $X_L = \omega L$. For example a 10 mH inductor fed with a 1 kHz signal has a reactance

$$X_L = 2\pi \times 1 \times 10^3 \times 10 \times 10^{-3} = 2\pi \times 10 = 62.8 \; \Omega$$

The reactance of an inductor increases with frequency (Fig. 4.26).

A capacitor has a capacitive reactance X_C,

$$X_C = \frac{1}{2\pi f C} = \frac{1}{\omega C}$$

where C is the capacitance in farads. A 1 μF capacitor, for example, operating at 10 kHz has a reactance

$$X_C = \frac{1}{2\pi \times 10 \times 10^3 \times 1 \times 10^{-6}} = \frac{1}{2\pi \times 10^{-2}}$$

$$= \frac{10^2}{2\pi} = 15.9 \; \Omega$$

The reactance of a capacitor decreases with increasing frequency (Fig. 4.27).

A network consisting of a capacitive reactance X_C and an inductive reactance X_L has a total reactance given by the phasor sum of X_C and X_L. The two vectors X_C and X_L are in anti-phase as shown in Fig. 4.28(b), i.e. they have a phase difference of 180°. Hence the total reactance is simply the difference between the values of X_C and X_L.

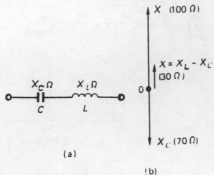

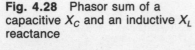

(a)

Fig. 4.28 Phasor sum of a capacitive X_C and an inductive X_L reactance

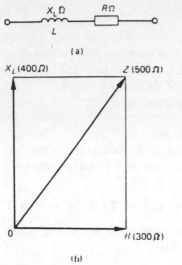

Fig. 4.29 (a) Inductor L in series with resistor R (b) Phasor representation of R, X_L and their phase sum Z

For example, given that $X_L = 100\ \Omega$ and $X_C = 70\ \Omega$, then the total reactance $X = 100 - 70 = 30\ \Omega$ which is inductive since X_L is greater than X_C.

Impedance

A network consisting of a reactance (inductive or capacitive) and a resistance has a total a.c. resistance known as its impedance Z. The impedance Z is the phasor sum of reactance X and resistance R.

For example, consider the reactive inductance X_L in series with resistor R in Fig. 4.29. As shown in 4.29(b), phasor X_L *leads* phasor R by 90° giving

Impedance $Z = \sqrt{(X_L^2 + R^2)}$

For $X_L = 400\ \Omega = 4 \times 10^2\ \Omega$ and $R = 300\ \Omega = 3 \times 10^2\ \Omega$,
$Z = \sqrt{(16 \times 10^4 + 9 \times 10^4)} = \sqrt{(25 \times 10^4)} = (5 \times 10^2)$
$= 500\ \Omega$

5 Amplifiers and oscillators

Amplifiers are very widely used devices in all types of electronic systems. They have the property of amplification, i.e. the process of producing an output larger than the original input.

Gain

The gain of an amplifier is the number of times its output is bigger than its input.

$$\text{Gain} = \frac{\text{Output}}{\text{Input}} \text{ and has no units — it is merely a ratio.}$$

There are three kinds of gain:

$$\text{Voltage gain } G_v = \frac{\text{Output voltage}}{\text{Input voltage}} = \frac{V_o}{V_{in}} \text{ (no units)}$$

$$\text{Current gain } G_i = \frac{\text{Output current}}{\text{Input current}} = \frac{I_o}{I_{in}} \text{ (no units)}$$

$$\text{Power gain } G_p = \frac{\text{Output power}}{\text{Input power}} = \frac{P_o}{P_{in}} \text{ (no units)}$$

Gain of two or more stages

A 2-stage amplifier is as shown in Fig. 5.1.

The output V_1 from the first stage is fed into the second stage. Thus V_1 becomes the input to the second stage. The overall input is

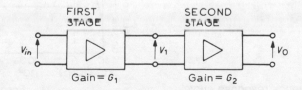

Fig. 5.1 Gain of two stages = $G_1 \times G_2$

V_{in} and the overall output is V_o. *The overall gain G of the amplifier is given as*

$$G = \frac{\text{Overall output}}{\text{Overall input}} = \frac{V_o}{V_{in}}$$

Assuming the gain of the first stage G_1 is 20 and that of the second stage G_2 is 50, then if the input voltage, V_{in} was 1 mV,

V_1 = gain of the 1st stage $\times$ V_{in} = 20 $\times$ 1 mV = 20 mV
V_o = gain of the 2nd stage $\times$ V_1 = 50 $\times$ 20 mV
$\qquad\qquad\qquad\qquad\qquad\quad$ = 1000 mV = 1 V

Overall gain $G = \dfrac{V_o}{V_{in}} = \dfrac{1000 \text{ mV}}{1 \text{ mV}} = 1000$

The overall gain of the amplifier can more easily be calculated as

G = Gain of first stage $\times$ Gain of second stage
$\quad$ = $G_1 \times G_2$ = 20 $\times$ 50 = 1000

Similarly, the gain for a 3-stage amplifier is given by

$$G = G_1 \times G_2 \times G_3$$

and so on for any number of stages.

Frequency response

The frequency response curve of an amplifier shows how the gain or the output of the amplifier changes with the signal frequency.

The frequency response curve (Fig. 5.2) is the graph of voltage output (or gain) on the y-axis against frequency on the x-axis. A logarithmic (non-linear) scale is used for the frequency — this spreads out the low-frequency end and compresses the high-frequency end.

The output (and the gain) of the amplifier is constant at the mid-frequency range but falls at high and low frequencies. The range of frequencies between f_1 and f_2 is the bandwidth of the amplifier. f_1

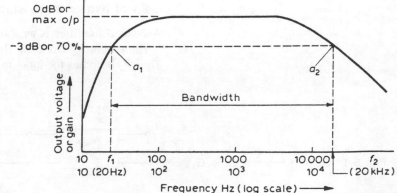

Fig. 5.2 Frequency response curve

and f_2 correspond to points a_1 and a_2 on the curve known as the 3 dB (decibel) points. The decibel is a non-linear (logarithmic) unit for measuring gain (see Appendix II). Point a_1 is known as the lower 3 dB point while a_2 is known as the upper 3 dB point. These two points are chosen where the voltage output of the amplifier has fallen to 70% of its maximum value. The 3 dB points are also known as the half-power points since the output power of the amplifier at the corresponding two frequencies is half its maximum value.

The output voltage (or the gain) on the y-axis can be represented in decibels, the maximum as 0 dB and the level for the 3 dB points as -3 dB.

The a.f. response curve

In order to amplify all audio frequencies the a.f. amplifier must have a bandwidth equivalent to the a.f. range, i.e. from 20 Hz to 20 kHz. Along this range of frequencies the gain of the amplifier should be steady. Below 20 Hz and above 20 kHz, the gain falls. Figure 5.2 shows a typical a.f. response curve.

As music and speech contain a complex mixture of frequencies within the a.f. band, the quality of an amplifier depends on how many of these frequencies and their harmonics it can reproduce without distortion. A small bandwidth will necessarily restrict the number of harmonics that can be reproduced at the output, hence the low quality of sound produced by cheap amplifiers.

There are two main types of distortion: amplitude and frequency.

Amplitude–amplitude distortion

Every amplifier has a maximum output that it cannot exceed. Once exceeded, the output will suffer from amplitude distortion. This is shown by the flattening or cutting off of one or both peaks of the waveform.

Figure 5.3 shows three ways in which a sine wave input can suffer amplitude distortion at the output of an overdriven amplifier.

Alternation–frequency distortion

Amplifiers are required to reproduce the input waveform without any change in its shape apart from an increase in its amplitude. Usually, the input signal is a complex wave consisting of a number of sine waves of different frequencies and their harmonics. For a faithful reproduction, all these frequencies must be amplified by the same ratio. That is to say the gain of the amplifier must be the same for all of them. In other words, the frequency response curve of the amplifier must be reasonably flat over the whole bandwidth, otherwise the output will suffer from frequency distortion.

For example, an amplifier with the response curve shown in Fig. 5.4

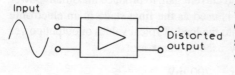

Input

Distorted output

(a)   (b)   (c)

Fig. 5.3   Amplitude distortion

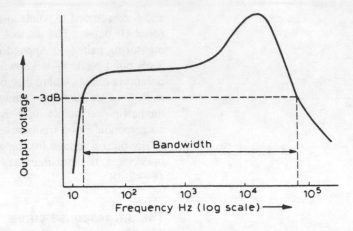

Fig. 5.4

will have the high frequencies amplified more than the lower frequencies. The output will contain an over-emphasis of high frequencies. The output is then said to suffer from frequency distortion.

Voltage amplifiers

These are the most commonly used amplifiers. They have a large voltage gain to provide maximum voltage at the output. They are used whenever a large voltage swing is required, e.g. in a driver stage to drive the power amplifier.

Power amplifiers

These have a large power and current gain to produce maximum power output. A power amplifier is used as the final stage in an electronic system to drive the power into the load. Typical values of output power requirements of commonly used systems are:

small radio amplifier	200 mW
hi fi	10 W
public address system	100 W or more

Push—pull amplifier

In order to get a high power output from an amplifier without amplitude distortion, a push—pull power amplifier is used (Fig. 5.5).

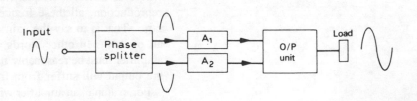

Fig. 5.5 Push—pull amplifier

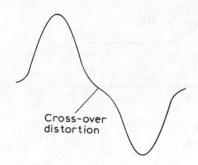

Fig. 5.6 Cross-over distortion

In the push–pull amplifier, the signal is first fed into a phase splitter which splits the waveform into two halves: a positive half and a negative half. Each half is then amplified by a separate amplifier, amplifier A_1 for the positive half and amplifier A_2 for the negative half. Once amplified, the two halves are added together by the mixer or output unit to reproduce the complete cycle which is fed into the loudspeaker load.

The output of the push–pull amplifier suffers from a distortion in the waveform known as cross-over distortion shown in Fig. 5.6. Cross-over distortion is easily overcome by proper design of the circuit.

Wide-band amplifier

A wide-band amplifier has a large bandwidth starting at practically zero Hz (d.c.) up to a few MHz. The wide bandwidth is achieved at the expense of the gain. Compared with the a.f. amplifier, the wide-band amplifier has a much larger bandwidth but a lower gain.

Wide-band amplifiers are used in TV receivers and for pulse amplification where a large band of frequencies is present — up to 5 MHz or more (Fig. 5.7). These amplifiers are also known as video amplifiers. Other uses are in radar and as Y-amplifiers for oscilloscopes.

A wide-band amplifier is also used for a complex wave with a large number of harmonics. The greater the number of harmonics present, the wider the bandwidth of the amplifier must be, otherwise distortion occurs. For a square wave with an infinite number of harmonics, the bandwidth required is also infinite. However, in practice the higher harmonics, the ninth or eleventh onwards, being small in amplitude, are neglected without much damage to the waveform.

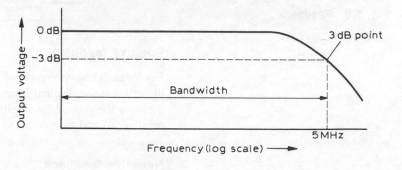

Fig. 5.7 Wide-band video amplifier

Radio frequency amplifiers

Radio frequency amplifiers are used in radio transmitters and receivers. They are selective amplifiers tuned to one particular radio frequency, and have a very narrow bandwidth and a very high gain.

A typical response curve is shown in Fig. 5.8. It shows the narrow

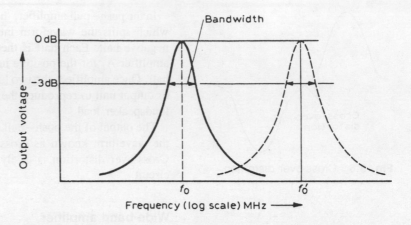

Fig. 5.8 R.F. amplifier response curve

bandwidth between the two 3 dB points. f_0 is the tuning frequency of the amplifier. When tuned to a different frequency f_0', the response curve shifts along to the position shown by a dotted line.

Feedback

Feedback is the process of feeding part or all of the output of an amplifier back to the input (Fig. 5.9). There are two types of feedback: positive and negative.

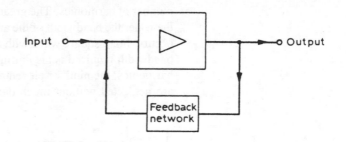

Fig. 5.9 Feedback

Positive feedback

Feedback is known as positive when the signal fed back is in the same direction as the original input. The effect of positive feedback is to *increase* the gain of the amplifier (leading to instability and possible oscillation) and to reduce the bandwidth.

Negative feedback

Feedback is known as negative when the signal fed back is in opposition to the original input. Contrary to positive feedback, this kind of feedback *reduces* gain, improves stability and increases the bandwidth of the amplifier. The effect of negative feedback on the response curve is to make it more steady and flat as shown in Fig. 5.10.

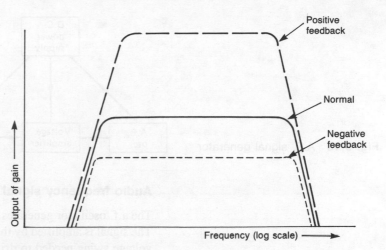

Fig. 5.10 Effect of feedback on the frequency response curve

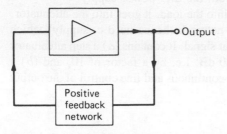

Fig. 5.11 The oscillator

Oscillators

These are amplifiers with such a large amount of positive feedback that an output is produced without any 'external' input (Fig. 5.11). The level of the output is determined by the gain of the amplifier as well as the feedback network.

As explained in Chaper 32 sine wave oscillators use tuned circuits to select the frequency of oscillation. The frequency of oscillation can be changed by changing the component values of the tuned circuit.

Square wave oscillators use no tuned circuits. Instead an *RC* feedback network is used, e.g. as in a multivibrator. The shape, i.e. the mark-to-space ratio, and the frequency of the square wave are determined by the values of *R* and *C* used.

One important requirement of an oscillator is that its frequency should remain constant while it is in use. Changes in the frequency of oscillation occur for two reasons. First, temperature change may cause component values to drift, and these changes produce **frequency drift**. Frequency drift can be overcome by the use of compensatory circuits to counterbalance these changes keeping the frequency constant at all temperatures, or keeping the frequency determining units in a constant temperature oven.

Second, change may occur due to incorrect loading of the oscillator. This is known as **frequency pulling**. To prevent this happening, a buffer stage is used to isolate the oscillator from the load. The load could be a loudspeaker or an amplifier stage (as shown in Fig. 5.12). The buffer stage presents the correct load to the oscillator no matter what the final load is.

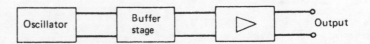

Fig. 5.12 The use of a buffer stage

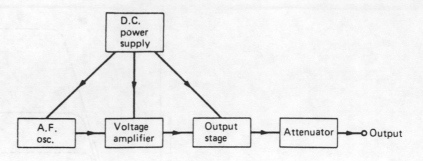

Fig. 5.13 A.F. signal generator

Audio frequency signal generator (Fig. 5.13)

The a.f. oscillator generates the signal and determines its frequency. The signal is amplified by the voltage amplifier to give the necessary voltage swing needed to drive the output stage. The output stage is a power amplifier which drives the power into the load. The a.f. oscillator, voltage amplifier and the power output stage are fed with the necessary d.c. power from the d.c. power supply.

Before feeding the signal into the load, it goes into the attenuator. The attenuator is a passive network (hence no d.c. supply) which controls the level of the output signal. It contains (*a*) a step attenuator giving control in steps of 20 dB, i.e. by a factor of 10, and (*b*) a continuous attenuator giving continuous and fine control of the output level.

6 Resonance

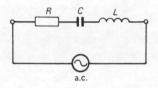

Fig. 6.1 Series resonance

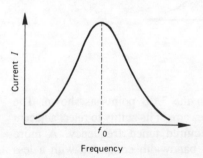

Fig. 6.2 Response curve for a series resonant circuit

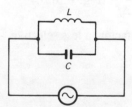

Fig. 6.3 Parallel resonance

Series resonance

Consider the series network in Fig. 6.1 with resistor R in series with inductor L and capacitor C. Resistor R may be an actual resistor or may represent the resistance of the inductor. The value of R is independent of frequency and stays constant at all frequencies. The reactances of L and C, on the other hand, are frequency-dependent. Thus as the frequency is increased from d.c. (zero Hz), the reactance of L increases while that of C decreases until at one frequency f_0, both reactances X_L and X_C become equal. At this point the circuit is said to be at resonance, with f_0 as the frequency of resonance.

Being in anti-phase to each other X_L and X_C cancel each other, resulting in a purely resistive impedance $Z = R$. With Z at this minimum value, the current through the circuit is at maximum, giving the response curve shown in Fig. 6.2. Because a series resonant circuit has a maximum current at resonance, it is also known as an **acceptor** circuit.

Parallel resonance

Resonance may be obtained with the parallel circuit shown in Fig. 6.3. At one specified frequency determined by the values of L and C the reactance of L cancels that of C and produces resonance. The impedance is now at a maximum giving maximum voltage. The circuit is thus known as a **rejector**, with a frequency response shown in Fig. 6.4.

For both series and parallel resonance, the resonant frequency is given as

$$f_0 = \frac{1}{2\pi\sqrt{(LC)}}$$

Given that $L = 10$ mH and $C = 1$ μF then

$$f_0 = \frac{1}{2\pi \times \sqrt{(10 \times 10^{-3} \times 1 \times 10^{-6})}}$$

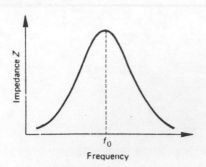

Fig. 6.4 Response curve for a parallel resonant circuit

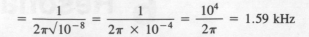

$$= \frac{1}{2\pi\sqrt{10^{-8}}} = \frac{1}{2\pi \times 10^{-4}} = \frac{10^4}{2\pi} = 1.59 \text{ kHz}$$

Note that the value of the resistance in the circuit does not affect the frequency at which the resonance occurs.

Bandwidth and selectivity

Resonant circuits are used mainly as tuned circuits because of their selective frequency response. Normally, parallel tuned circuits are used since they have a high impedance and thus a high voltage output.

The frequency response of a tuned circuit is shown in Fig. 6.5.

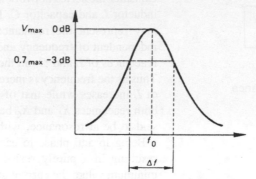

Fig. 6.5 Frequency response curve for a tuned circuit. Δf is the bandwidth

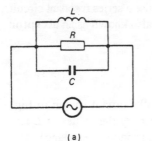

(a)

The bandwidth is taken between the 3 dB points as shown. The selectivity of a tuned circuit is a measure of its ability to reject adjacent frequencies in favour of the required tuned frequency. A more selective circuit has a narrower bandwidth compared with a less selective circuit having a wider bandwidth. Selectivity is measured by its Q-factor which is defined as

$$\text{Q-factor} = \frac{\text{Resonant frequency}}{\text{Bandwidth}} = \frac{f_0}{\Delta f} \text{ (Fig. 6.5)}$$

Hence a highly selective circuit has a high Q-factor; a less selective circuit has a low Q-factor.

Damping

Where a wide bandwidth is required, say for a TV receiver, low-Q tuned circuits are needed. The low Q-factor may be achieved by shunting the tuned circuit with a resistor R as shown in Fig. 6.6(a). The resistor has what is known as a damping effect on the response curve, making the circuit less selective, as shown in 6.6(b).

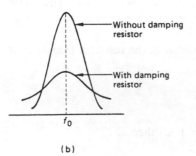

(b)

Fig. 6.6 Damping effect on a tuned circuit

Damped oscillation

Consider a resonant circuit tuned to a frequency of 1 kHz. Since the circuit rejects all other frequencies, the only output that may appear across it will be in the form of a 1 kHz sine wave. Such an output may be produced by feeding the circuit with a.c. energy which makes it oscillate sinusoidally. This a.c. energy may be in the form of a sine wave at the resonant frequency or a complex waveform having a harmonic at the resonant frequency.

Figure 6.7 shows a resonant circuit tuned to 3 kHz fed with a 1 kHz square wave. The tuned circuit will oscillate to the third harmonic of the square wave (i.e. 3 kHz) producing sustained oscillations as shown. Figure 6.8 shows the same tuned circuit, this time fed with a step voltage generated by closing switch S. The step voltage contains an infinite number of harmonics which energises the tuned circuit into oscillating at its 3 kHz resonant frequency. However, the oscillations are damped, i.e. dying out, since the a.c. energy produced by the step voltage is gradually lost due to the small losses in L and C as well as in the resistance of the connecting wires, etc. After a while the oscillations stop completely. Another set of damped oscillations may be produced by opening switch S.

Damped oscillation occurs whenever the energy fed into the resonant circuit is not high enough or frequent enough to overcome the losses in the circuit.

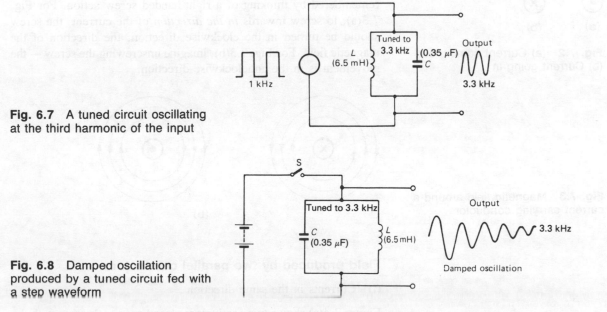

Fig. 6.7 A tuned circuit oscillating at the third harmonic of the input

Fig. 6.8 Damped oscillation produced by a tuned circuit fed with a step waveform

7 Transformers

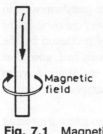

Fig. 7.1 Magnetic field due to current

(a)   (b)

Fig. 7.2 (a) Current coming out (b) Current going in

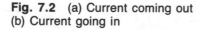

Fig. 7.3 Magnetic field around a current carrying conductor

Electromagnetism

When a current flows through a conductor, a magnetic field is produced around it (Fig. 7.1). The direction of the magnetic field is determined by the direction of the current.

The notation for the direction of current through a conductor is as shown in Fig. 7.2. The point in (a) can be thought of as the pointed end of an arrow or dart, while the cross in (b) can be thought of as the tail end view of the arrow or dart.

The magnetic field surrounding a conductor carrying a current is as shown in Fig. 7.3. The direction of the field can easily be remembered by thinking of a right-handed screw action. For Fig. 7.3(a), to screw inwards *in the direction* of the current, the screw should be turned in the clockwise direction, the direction of the magnetic field. For Fig. 7.3(b), imagine unscrewing the screw — the movement is in the anticlockwise direction.

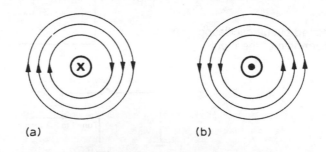

(a)   (b)

Field produced by two parallel conductors

(i)   Currents in the same direction

Figure 7.4(a) shows two conductors placed near each other with the magnetic field of each conductor drawn separately. The two fields oppose each other and cancel out between the conductors. The resultant is a single field surrounding both conductors as shown in

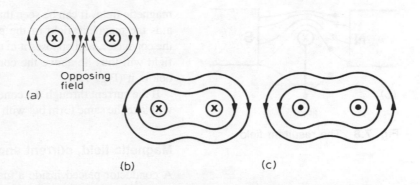

Fig. 7.4 (a) Two conductors with current going in the same direction (b) and (c) show resultant field

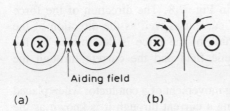

Fig. 7.5 (a) Two conductors with current going in opposite directions (b) The resultant field

(b). If the direction of the currents is reversed, the direction of the magnetic field will also be reversed as shown in (c).

(ii)   Currents in opposite direction

Figure 7.5(a) shows the magnetic field of each conductor separately. This time the two fields are aiding each other in the middle. The resultant field is as shown in (b) with the magnetic field strongest between the conductors.

Field produced by a solenoid

A solenoid is a coil of wire as shown in Fig. 7.6. When current passes through it, the solenoid behaves like a magnetic bar with north and south poles. Its magnetic field is the same as that for a permanent magnet. The magnetic field can be strengthened by winding the solenoid coil round a magnetic core such as iron or steel. The current and the number of turns also determine the strength of the magnetic field.

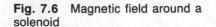

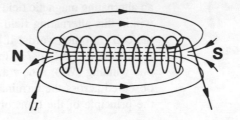

Fig. 7.6 Magnetic field around a solenoid

Electromagnet

One use of the solenoid is as an electrogmagnet. A soft magnetic material such as soft iron is used as the core. The solenoid behaves as a magnet only when there is current flowing through the coil. Electromagnets are used in electric bells and relays.

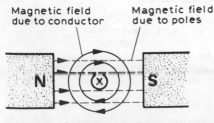

Magnetic field due to conductor   Magnetic field due to poles

Fig. 7.7 Conductor inside a magnet field

A conductor inside a magnetic field

Figure 7.7 shows a conductor carrying a current placed inside a

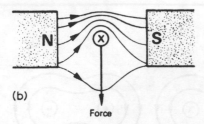

(b)

Fig. 7.8 The resultant field

magnetic field. It can be seen that the magnetic field of the conductor aids the magnetic field of the permanent magnet in the area above the conductor, and opposes it in the area below it. A stronger magnetic field will thus lie above the conductor whilst a weaker one will lie below it (Fig. 7.8).

If the current through the conductor is reversed, the magnetic field will take the same form but with the strong field below the conductor.

Magnetic field, current and movement

A conductor placed inside a magnetic field will experience a force which tries to move it away from the strong magnetic field towards the weaker field as shown in Fig. 7.8. The direction of the force depends on the direction of current as well as the direction of the magnetic field. The magnitude of the force is determined by the strength of both the magnetic field and the current through the conductor.

The principle of producing movement of a conductor when placed in a magnetic field by passing a current through it is known as the **motor** principle and is used in electric motors, moving coil meters, and other devices.

If a conductor is moved across a magnetic field, a current is generated in the conductor. This is known as the **generator** principle and is used in dynamos and alternators.

Induced e.m.f.

So far, only the magnetic field due to direct current has been considered. The direction of the magnetic field is determined by the direction of the d.c. current. When an alternating current is used, an alternating magnetic field is set up. If a separate coil is now placed within this alternating field, an e.m.f. (voltage) will be induced in the coil. In other words, if two separate coils are placed near each other as shown in Fig. 7.9, then when an a.c. voltage is applied across one winding (W_1), a new a.c. voltage (the induced e.m.f.) will be observed across the terminals of the second winding (W_2). This is the principle of the transformer.

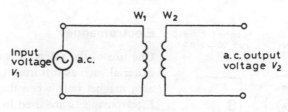

Fig. 7.9 Induced e.m.f.

The transformer

The transformer consists of two separate windings known as the *primary* and *secondary* windings. The input (a.c.) voltage is applied across the primary, producing a changing magnetic field. This changing magnetic field interacts with the secondary winding inducing an a.c. voltage across it. The induced voltage at the secondary has the same frequency as the input voltage but with its amplitude determined by the number of turns of both the primary and secondary windings.

If the input voltage at the primary $= V_1$
the output voltage at the secondary $= V_2$
the number of turns of the primary $= T_1$
the number of turns of the secondary $= T_2$ then

$$\frac{V_1}{V_2} = \frac{T_1}{T_2}$$

Also, $\dfrac{I_1}{I_2} = \dfrac{T_2}{T_1}$ where I_1 and I_2 are primary and secondary currents.

Transformer efficiency

The above ratios assume that a transformer is 100% efficient, i.e. having no power loss. Hence

Input power $I_1 \times V_1$ = Output power $I_2 \times V_2$

In practice, transformers have efficiencies ranging from 96 to 99%.
   To improve the efficiency of the transformer, both primary and secondary windings are wound on one magnetic core as shown in Fig. 7.10.

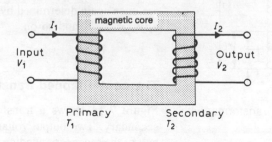

Fig. 7.10 The transformer

Step-up and step-down

A step-up transformer produces a higher voltage at the output (across its secondary) than the input voltage (across its primary). To achieve this the number of turns on the secondary has to be larger than those on the primary.
   A step-down transformer produces a smaller voltage at its output because it has a smaller number of turns on its secondary than on its primary.

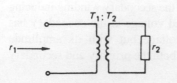

Fig. 7.11 Impedance ratio $r_1/r_2 = T_1^2/T_2^2 = n^2$

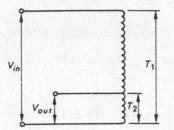

Fig. 7.12 The auto-transformer

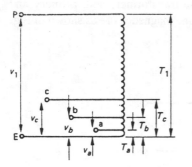

Fig. 7.13 An auto-transformer with multiple tappings

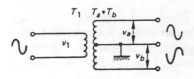

Fig. 7.14 The centre tapped transformer

Impedance ratio

Consider the transformer shown in Fig. 7.11 with a load resistor r_2 connected across its secondary. Looking into the primary, an input resistance r_1 is observed which is the equivalent value of r_2 when transferred to the primary. The ratio r_1/r_2 is known as the resistance or impedance ratio of the transformer. This ratio may be derived as follows. Since $r_1 = V_1/I_1$ and $r_2 = V_2/I_2$, then

$$\frac{r_1}{r_2} = \frac{V_1}{I_1} \div \frac{V_2}{I_2} = \frac{V_1}{I_1} \times \frac{I_2}{V_2} = \frac{V_1}{V_2} \times \frac{I_2}{I_1}$$

But $V_1/V_2 = T_1/T_2 = n$ and $I_2/I_1 = T_1/T_2 = n$, therefore

$$\frac{r_1}{r_2} = n \times n = n^2$$

For example, if $r_2 = 100\ \Omega$ and the turns ratio $T_1/T_2 = n = 2{:}1$, then looking into the primary winding, the transformer will be seen as a resistor $r_1 = 100\ \Omega \times 2^2 = 100 \times 4 = 400\ \Omega$.

Auto-transformer

A transformer may be made with a single winding employing one tapping as shown in Fig. 7.12. T_1 is the number of turns of the primary and T_2 is the number of turns of the secondary. The voltage, current, impedance, and turns ratios have the same relationship as for an ordinary transformer.

Figure 7.13 shows another single winding transformer employing more than one tapping. The voltage, current, and impedance ratios are once again determined by the turns ratio, e.g. $v_1/v_a = T_1/T_a$ and $v_1/v_b = T_1/T_b$ and so on.

The centre tapped transformer

Figure 7.14 shows a transformer with a tap at the centre of its secondary. Two output voltages v_a and v_b are produced across each half of the secondary winding. The ratio of the input (primary) voltage to the secondary voltages is given by the turns ratio whereby

$$\frac{v_1}{v_a} = \frac{T_1}{T_a} \quad \text{and} \quad \frac{v_1}{v_b} = \frac{T_1}{T_b}$$

where T_1, T_a and T_b are the number of turns of the primary, secondary a and secondary b respectively. Since the tapping is at the centre, v_a and v_b have equal amplitude. If the centre is earthed, as shown in Fig. 7.14, the two output waveforms are then in anti-phase to each other.

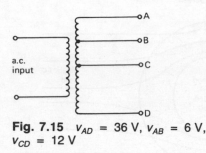

Fig. 7.15 v_{AD} = 36 V, v_{AB} = 6 V, v_{CD} = 12 V

EXAMPLE

Refer to Fig. 7.15. (*a*) Calculate the voltage between transformer tappings B and C. (*b*) If there are 30 turns between tappings A and B, how many turns will the whole secondary winding have?

Solution

(*a*)   Voltage between B and C = $v_{BC} = v_{AD} - v_{AB} - v_{CD}$.
$$= 36 \text{ V} - 6 \text{ V} - 12 \text{ V} = 18 \text{ V}.$$

(*b*) $\dfrac{v_{AB}}{v_{AD}} = \dfrac{\text{Number of turns between A and B}}{\text{Number of turns between A and D}}$

Hence $\dfrac{6 \text{ V}}{36 \text{ V}} = \dfrac{30}{T_{AD}}$ or $T_{AD} = 30 \times \dfrac{36}{6} = 180$ turns

The magnetic circuit

It is accepted that, in a magnetic circuit, the magnetic flux (or magnetic field) measured in Teslas is created by a force known as the magnetomotive force (m.m.f.). A magnetic circuit is usually compared with an electric circuit, where the flux corresponds to the current and the magnetomotive force m.m.f. corresponds to the electromotive force e.m.f. Just as an electric circuit is said to have a resistance, so a magnetic circuit is said to have a **reluctance** *S* which corresponds to resistance *R*. For example, a magnetic material such as soft iron has a low reluctance, i.e. low resistance to magnetic flux.

Permeability

The permeability of a material is a measure of the ease with which it may be magnetised. For example, soft iron and other electro-magnetic materials such as ferrites have high permeability. These materials are used in transformers, inductors, relays, and in ferrite aerials. On the other hand, non-magnetic materials have low permeability. Magnetic alloys such as silicon steel have the ability to retain their magnetism and are therefore used as permanent magnets in loudspeakers, moving-coil instruments, etc.

Screening

Consider the hollow cylinder, shown in Fig. 7.16, placed inside a magnetic field. If the cylinder is made of a low reluctance magnetic material, then the magnetic field will concentrate through the cylinder as shown, with no magnetic flux passing through the space inside the cylinder. Consequently a body placed in this space would be screened or shielded from the magnetic field around it. Such screening, known as magnetic screening, is used to protect devices such as a cathode ray tube, a moving-coil meter, or a loudspeaker from outside magnetic fields.

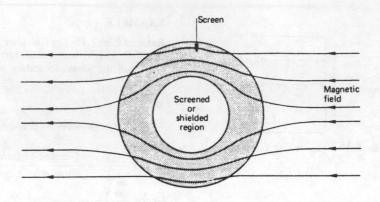

Fig. 7.16 Magnetic screening

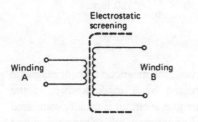

Fig. 7.17 Electrostatic screening as used in a transformer

In transformers, another type of screening known as electrostatic or electric screening is sometimes used. A screen or shield of thin copper sheet is placed between the primary and secondary windings as shown in Fig. 7.17. By earthing the screen, the effect of the capacitance that exists between the two windings due to their different potentials is minimized. Electrostatic screening is also used in coaxial cables and whenever conductors having different potentials are placed in close proximity.

8 Matching

Input and output impedance

Any device such as a simple amplifier or a complete TV receiver may be represented by a box with two input leads and two output leads (Fig. 8.1) and having an input impedance Z_{in} and an output impedance Z_o. Now, Z_{in} is the impedance of the device when looking into the input terminals AB, and Z_o is the impedance looking back into the output terminals CD. As far as the input signal is concerned, the device is a load impedance of Z_{in}. As far as the next stage of a load is concerned, the device looks like a generator with an internal impedance of Z_o.

In many cases, e.g. in power amplifiers or a receiving aerial feeding a radio or TV receiver, we wish to transfer power from one stage to the other. In these cases it is important that the maximum possible power is transferred. This is the purpose of impedance matching or maximum power transfer. Matching or maximum power transfer occurs when the output impedance of one stage is the same as the input impedance of the succeeding stage (Fig. 8.2).

As an illustration, consider the circuit in Fig. 8.3 where a generator having an internal impedance of $r = 10\ \Omega$ and an e.m.f. $e = 3$ V is used to feed a load resistance R. Maximum power transfer from the generator to the load takes place when load R = internal (or output) resistance of the generator.

For $R = 10\ \Omega$, voltage across R, $V_R = 3/2 = 1.5$ V. Hence output power or power transferred into the load is

$$\frac{V^2}{R} = \frac{1.5 \times 1.5}{10} = \frac{2.25}{10} = 0.225 \text{ W} = 225 \text{ mW}.$$

Note that the power dissipated into the load is the same as that dissipated in the internal resistance r when the load is correctly matched.

For any other value of load R, the output power will be smaller than 225 mW. For $R = 20\ \Omega$

$$V_R = \frac{3}{10 + 20} \times 20 = \frac{3 \times 20}{30} = 2 \text{ V}$$

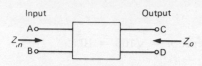

Fig. 8.1 Input and output impedances Z_{in} and Z_o

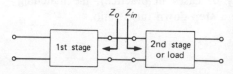

Fig. 8.2 For matching $Z_o = Z_{in}$

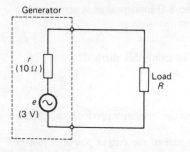

Fig. 8.3

giving an output power of

$$\frac{V^2}{R} = \frac{2 \times 2}{20} = \frac{4}{20} = 0.2 \text{ W} = 200 \text{ mW}.$$

Similarly if load R were reduced in value to say 10/2 or 5 Ω, then output power will be reduced to 200 mW.

Transformer matching

To ensure matching between two stages or between a load and an output stage in a circuit, a transformer is sometimes used as the *coupling device*. Figure 8.4 shows an a.f. power amplifier feeding into a loudspeaker load. Suppose that a 10 Ω loudspeaker is used with the amplifier having an output impedance of 1000 Ω. For matching to occur, the turns ratio $T_1/T_2 = n$ must be of such a value as to present 1000 Ω to the amplifier.

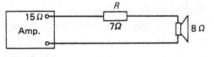

Fig. 8.4 Matching transformer. Turns ratio 10:1

But $n^2 = \dfrac{r_1}{r_2} = \dfrac{1000}{10} = 100$, hence $n = \sqrt{100} = 10$.

In other words, a step-down transformer must be used having a turns ratio of 10:1.

Note that in all cases where the amplifier is feeding a low impedance load (which is the majority of cases in practice), the matching transformer used must have a **step-down** turns ratio.

Examples

EXAMPLE 1

Figure 8.5 shows an amplifier feeding an 8 Ω loudspeaker in series with a 7 Ω resistor.

    (*a*) State the purpose of R.
    (*b*) What is the main disadvantage of using this method?

Solution

    (*a*) The purpose of R is to ensure that the amplifier produces maximum power output.
    (*b*) The disadvantage is that almost half of the output power is lost in the resistor.

EXAMPLE 2

Figure 8.6 shows a method used to provide a facility for earphone connection. State the value of R if the earphone has an impedance of 8 Ω.

Solution

$R = 30 - 8 = 22$ Ω.

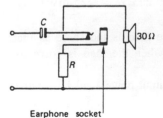

Fig. 8.5

Fig. 8.6

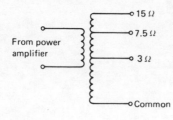

Fig. 8.7

EXAMPLE 3

Refer to Fig. 8.7. How should two 15 Ω speakers be connected to the output transformer shown?

Solution

For matching to take place, output impedance presented by the transformer must be equal to the load resistance. This can be done by connecting the two speakers in parallel, thus giving a total load of 7.5 Ω across the common connection and the transformer tapping labelled 7.5 Ω

EXAMPLE 4

Figure 8.8 shows a signal splitter network where an extra load is matched to the same source. Calculate the value of *R*.

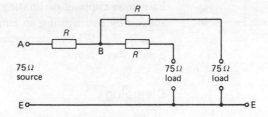

Fig. 8.8

Solution

For matching the total resistance R_{AE} looking into AE must be 75 Ω. The total resistance R_{BE} between B and E (across the two load branches) is

$$\frac{1}{2}(R+75) = \frac{R}{2} + \frac{75}{2}$$

Total resistance across AE is

$$R_{AE} = R + R_{BE} = R + \frac{R}{2} + \frac{75}{2} = \frac{3R}{2} + \frac{75}{2}$$

But for matching, $R_{AE} = 75$ Ω. Therefore

$$75 = \frac{3R}{2} + \frac{75}{2}$$

i.e. $\quad \frac{3R}{2} = 75 - \frac{75}{2} = \frac{75}{2}$

$$R = \frac{75}{3} = 25 \text{ Ω}$$

9 Wave shaping

Clipping

The process of flattening one or both peaks of an a.c. waveform is known as clipping or limiting. It is achieved as we saw in the last section by overloading an amplifier. A large degree of clipping as seen in Fig. 9.1(c) can produce an approximate square wave from a sine wave input.

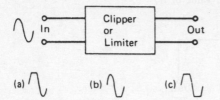

Fig. 9.1 Clipping

Clamping

A clamper is also known as a d.c. restorer. It provides an a.c. signal with a d.c. level which it did not have before. For example, clampers are used in TV receivers to give the video signal a d.c. level (Fig. 9.2).

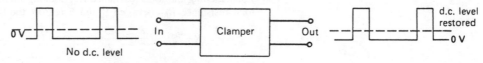

Fig. 9.2 Clamping

A.C. coupling

The reverse of the d.c. restorer is the a.c. coupler which removes the d.c. level from the input (Fig. 9.3). The a.c. coupler is usually as simple as a single capacitor blocking the d.c. while allowing the a.c. to go through.

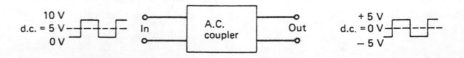

Fig. 9.3 A.C. coupling

Rectification

This is the process of transforming alternating current into direct current by allowing the current to flow in one direction only. Rectifiers are used in d.c. power supplies and for various other purposes such as demodulators.

Rectification is achieved in one of two ways.

(a) Half wave rectification which cuts out either the negative or the positive half of the signal. Only one half of the cycle is produced at the output as shown in Fig. 9.4. Such waveforms have a d.c. level equal to 0.318 (approximately $\frac{1}{3}$) of the peak value.

(b) Full wave rectification which reproduces the two halves of the cycle, but now both in the same direction (Fig. 9.5). The d.c. level is now doubled to 0.636 (approximately two-thirds) of the peak value.

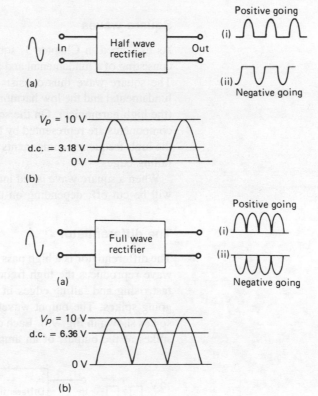

Fig. 9.4 Half wave rectification (b) shows d.c. level 0.318 V_p

Fig. 9.5 Full wave rectification (b) shows d.c. level = 0.636 V_p

Half wave rectification produces a signal having a frequency equal to the original wave while full wave rectification doubles the frequency of the original wave.

Filters

In electronics it is often necessary to separate one frequency band from another. A filter allows one frequency band through and cuts off all others. There are three main types of filter.

(a) The **high pass filter** allows only high frequencies to pass through and prevents all low frequencies. For a complex wave input containing a number of harmonics, only the high frequency harmonics will pass through while the fundamental and lower harmonics will be greatly attenuated.

(*b*) The **low pass filter** allows low frequencies to pass through and cuts off all high frequencies. For a complex wave input only the fundamental and low harmonics will appear at the output of this type of filter, while higher harmonics will be greatly attenuated.

(*c*) The **band pass filter** allows only a specific range or band of frequencies through and cuts off all other frequencies above and below that band.

Square waves

As explained in Chapter 3, square waves are complex waveforms consisting of a fundamental and an infinite number of odd harmonics. The square wave thus consists of low-frequency components (the fundamental and the low harmonics) and high-frequency components (the high harmonics). On the square wave itself, the low-frequency components are represented by the flat top and the flat bottom, while the high-frequency components are represented by the fast rising or falling edges.

When a square wave is fed into a filter, one of its two components will be cut off, depending on the type of filter used.

The differentiator

The differentiator is a high pass filter, which when fed with a square wave reproduces the high-frequency components only, namely the fast rising and falling edges in the form of positive- and negative-going spikes. The output waveform is the short duration pulses or spikes shown in Fig. 9.6. Each cycle of the input wave produces two spikes at the output, of an amplitude equal to that of the input.

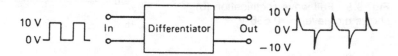

Fig. 9.6 Differentiation

The integrator

The integrator is a low pass filter which when fed with a square wave reproduces the low-frequency components only, namely the flat parts of the waveform. Ideally, only d.c. appears at the output. In practice, however, the output is a triangular wave with an amplitude smaller than that of the square-wave input (Fig. 9.7).

Fig. 9.7 Integration

The effect on a rectangular wave

Figure 9.8 shows the effect of differentiating and integrating a rectangular waveform.

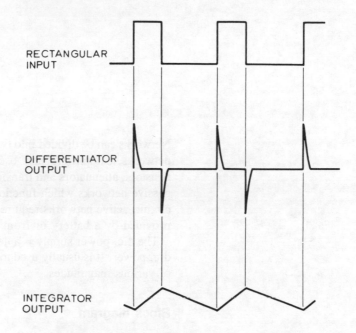

RECTANGULAR
INPUT

DIFFERENTIATOR
OUTPUT

INTEGRATOR
OUTPUT

Fig. 9.8 Differentiating and integrating a rectangular wave

The effect on a sine wave

A sine wave being a pure waveform will suffer no change in its shape when fed into any kind of filter such as a differentiator or an integrator. Only its amplitude will be affected. A differentiator being a high pass filter will, however, leave the amplitude of a high frequency sine wave unaffected. But a low frequency sine wave will suffer a reduction on its amplitude through a differentiator. The opposite is true for an integrator.

10 Power supplies I

Networks can be divided into two types, active networks and passive networks. Amplifiers and oscillators are active networks while resistors, attenuators and transformers are passive networks. Unlike passive networks which function merely by being connected in the circuit, active networks require d.c. power. This d.c. power can be provided by a battery or from a mains source.

The d.c. power supply is a device that transforms a.c. power into d.c. power. It is usually used to convert the mains into d.c. voltages of various magnitudes.

Block diagram

Figure 10.1 shows a block diagram for a d.c. power supply (*a*) using a half wave and (*b*) using a full wave rectifier. The a.c. input is usually taken from the mains. The first stage is the rectifier (either half wave or full wave rectifier). The output of the rectifier has a d.c. level

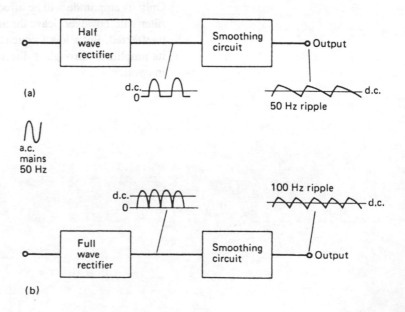

Fig. 10.1 D.C. power supply

56

together with a large a.c. variation. This waveform is known as a pulsating d.c. voltage which cannot be used as a satisfactory d.c. supply. In order to remove the a.c. content a smoothing circuit is used — which is a low pass filter allowing the d.c. to pass on and reducing the a.c. to a small ripple that appears at the output as shown. The frequency of the ripple is determined by the rectifier used. A half wave rectifier will produce a ripple frequency the same as the input, while full wave rectifiers double the frequency.

In many d.c. supplies, a transformer is used before the rectifier to change the a.c. input to the required voltage level (Fig. 10.2). The turns ratio of the transformer determines the d.c. output level produced.

Fig. 10.2 D.C. power supply using a transformer

Regulation curve

All d.c. supplies including batteries provide a maximum voltage under no-load condition, i.e. when the supply is providing no current. Due to internal resistance, the terminal voltage begins to fall below this maximum as load current is drawn from the supply. The way the terminal voltage changes with load current is known as the regulation of the power supply. A typical regulation curve is shown in Fig. 10.3.

In order to improve the regulation of a d.c. supply, i.e. to keep the terminal voltage constant as load current is increased, regulators are used at the output. Figure 10.4 shows a block diagram for a regulated power supply.

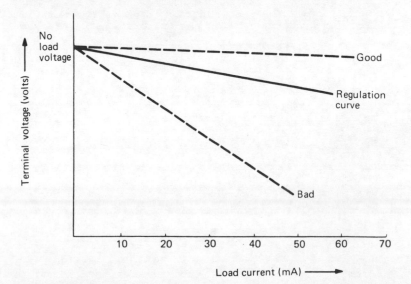

Fig. 10.3 Regulation curve

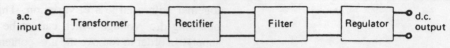

Fig. 10.4 Regulated power supply

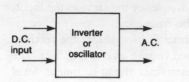

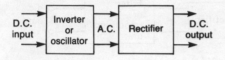

Fig. 10.5

Fig. 10.6

Inverters and converters

An inverter is a power supply which changes a d.c. input into an a.c. output (Fig. 10.5) while a converter changes a d.c. input voltage into another d.c. value. A converter is an inverter with a rectifier connected at the output to change the a.c. from the oscillator back into d.c. (Fig. 10.6).

11 Digital gates and system

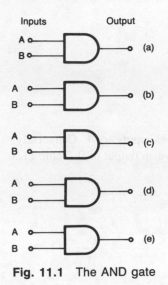

Inputs   Output

A   B   (a)

A   B   (b)

A   B   (c)

A   B   (d)

A   B   (e)

Fig. 11.1 The AND gate

Digital control is a relatively new area in electronics which developed from the science of telecommunication. Digital control embraces computers, electronic telephone exchanges, industrial control systems, data processing, and other similar systems. Digital devices function in discrete steps, usually ON or OFF. The essential requirement of a digital device is that it should change from one state to another without settling in any intermediate position.

Gates

Gates are logic or digital devices that give a 2-state output: a zero output (indicated by 0) and a fixed output (indicated by 1). A gate can have several inputs but produces only one output.

AND

The AND gate produces an output of 1 when all its inputs are 1. Figure 11.1 shows an AND gate with two inputs. Only when both input A and B are 1 does the output become 1. Table 11.1 represents the AND function in what is known as the **truth table** for the gate.

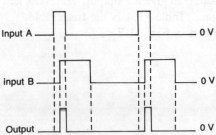
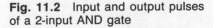

Fig. 11.2 Input and output pulses of a 2-input AND gate

Input A   0 V

input B   0 V

Output   0 V

Table 11.1 AND

Input		Output
A	B	
0	0	0
0	1	0
1	0	0
1	1	1

Typical input and output waveforms are shown in Fig. 11.2. Note that an output of 1 is produced only when both inputs are 1.

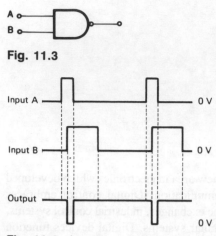

Fig. 11.3

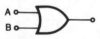

Fig. 11.4 Input and output pulses of a 2-input NAND gate

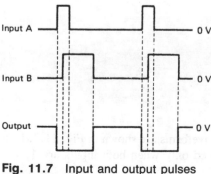

Fig. 11.5

Fig. 11.6

Fig. 11.7 Input and output pulses of a 2-input NOR gate

NAND (Fig. 11.3)

The NAND gate gives a 0 output when all its inputs are 1. Conversely, if a 0 is present at any of its inputs, the output will be 1. The NAND gate is thus the opposite of the AND gate. Table 11.2 is the truth table.

Typical input and output waveforms are shown in Fig. 11.4 in which a 0 output is produced when both inputs are 1.

Table 11.2 NAND

Input		Output
A	B	
0	0	1
0	1	1
1	0	1
1	1	0

OR (Fig. 11.5)

The OR gate gives a 1 output if any of its inputs are 1. Conversely, it requires a 0 on each input for it to give an output of 0. Table 11.3 is the truth table.

Table 11.3 OR

Input		Output
A	B	
0	0	0
0	1	1
1	0	1
1	1	1

NOR (Fig. 11.6)

The NOR gate gives a 0 output if any of its inputs is 1. Conversely it requires a 0 on each of its inputs to give a 1 output. The NOR gate is thus opposite to the OR gate. Table 11.4 is the truth table.

Typical waveforms are shown in Fig. 11.7

Table 11.4 NOR

Input		Output
A	B	
0	0	1
0	1	0
1	0	0
1	1	0

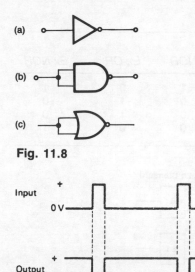

(a)

(b)

(c)

Fig. 11.8

Input

+

0 V

Output

+

0 V

Fig. 11.9 Input and output pulses of a NOT gate

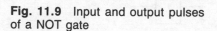

A

B

Fig. 11.10

A

B

Fig. 11.11

NOT (Fig. 11.8)

The NOT gate is an inverter. When the input is 1, the output is 0, and vice versa. Table 11.5 is the truth table.

An inverter may be produced by shorting the inputs of a NAND or a NOR gate, as shown in Fig. 11.8(b) and (c). Figure 11.9 shows typical input and output waveforms.

Table 11.5 NOT

Input	Output
0	1
1	0

Exclusive-OR (Fig. 11.10)

The Exclusive-OR (Ex-OR) gate gives an output of 1 when either of its inputs is 1 but not when both inputs are at logic 1 or logic 0. The truth table is shown in Table 11.6.

Table 11.6 Ex-OR

Input		Output
A	B	
0	0	0
0	1	1
1	0	1
1	1	0

Exclusive-NOR (Fig. 11.11)

The Exclusive-NOR (Ex-NOR) gate gives an output of 1 when its inputs are either both at 1 or both at 0. The truth table is shown in Table 11.7.

Table 11.7 Ex-NOR

Input		Output
A	B	
0	0	1
0	1	0
1	0	0
1	1	1

International and British standards

Figure 11.12 shows the international and British symbols for logic gates. A summary of the truth tables is shown in Table 11.8.

Table 11.8

Inputs A	B	AND	OR	NAND	NOR	Ex-OR	Ex-NOR
0	0	0	0	1	1	0	1
0	1	0	1	1	0	1	0
1	0	0	1	1	0	1	0
1	1	1	1	0	0	0	1

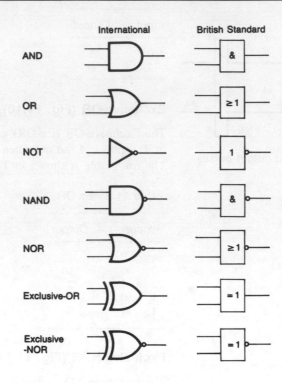

Fig. 11.12

Combinational gates

Digital systems usually involve more than one gate combined together to give the required control. For instance, a simple combination of an AND gate with a NOT gate (Fig. 11.13(a)) functions the same way as a NAND gate.

Similarly, a NOR gate can be constructed by an OR gate and a NOT gate as shown in Fig. 11.13(b).

As a further example, the input to the logic circuit shown in Fig. 11.14 is 0111; the question is to find the output.

Output of OR gate (i) is 1. Output of NAND gate (ii) is 0, giving an input to OR gate (iii) of 10 for an output at G of 1.

Figure 11.15 shows a simple gate combination used to control an alarm bell. Gate G_1 is a NOR gate with shorted inputs thus acting as a NOT gate. To operate the alarm, a positive voltage must be

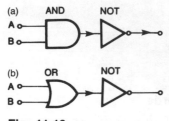

Fig. 11.13

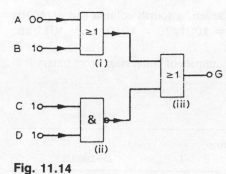

Fig. 11.14

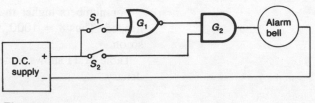

Fig. 11.15

applied to the bell, i.e. the output of the AND gate (G_2) must be at logic 1. For this to happen both inputs to the AND gate must also be at 1. It is therefore necessary for S_1 to be open and S_2 to be closed for the alarm to operate. No other combination of S_1 and S_2 can activate the alarm.

The binary system

While the decimal or denary system of counting uses ten different digits, the binary system uses two digits only, zero and one (0 and 1). This system fits in with the logic gates just dealt with.

In the denary system, the first column A (Table 11.9) is the ones or units column, column B is the tens, C is the hundreds, D is the thousands, and so on. In the binary system each column can either be 0 or 1 with the first column A corresponding to the ones, column B to the twos, column C to the fours, D to the eights, and so on. Any number can be written in the denary as well as the binary system. Table 11.10 shows the conversion into binary of the denary numbers nought to seven.

Table 11.9

Decimal columns				Binary columns			
D	C	B	A	D	C	B	A
10^3	10^2	10^1	10^0	2^3	2^2	2^1	2^0
1000s	100s	tens	units	eights	fours	twos	units

Table 11.10

Denary	Binary		
	C	B	A
	(4)	(2)	(1)
0	0	0	0
1	0	0	1
2	0	1	0
3	0	1	1
4	1	0	0
5	1	0	1
6	1	1	0
7	1	1	1

For numbers higher than seven, a fourth column (the eights) is needed. Thus $8 = 1000$, $9 = 1001$, $10 = 1010$, $11 = 1011$ and so on.

Table 11.11 shows some examples of conversion from binary into denary.

Table 11.11

Binary	Binary columns						Decimals
	32	16	8	4	2	1	
1110	—	—	1	1	1	0	$= 8 + 4 + 2 = 14$
1011	—	—	1	0	1	1	$= 8 + 2 + 1 = 11$
11001	—	1	1	0	0	1	$= 16 + 8 + 1 = 25$
10111	—	1	0	1	1	1	$= 16 + 4 + 2 + 1 = 23$
110010	1	1	0	0	1	0	$= 32 + 16 + 2 = 50$

Frequency counter

An AND gate in conjunction with a counter can be used to measure frequency or time period. Figure 11.16 shows a 2-input AND gate together with a counter used as a frequency meter. Input A is fed with the input signal. Input B is fed with a timing pulse of predetermined duration. At the output of the gate, the signal pulses will be present only for the duration of the timing pulse. These pulses are then fed into the counter which counts them, giving the frequency of the input signal. For example, if the timing pulse was of 10 ms duration and the frequency of the input was 1 kHz giving a periodic time of 1 ms, then only 10 pulses will appear at the output of the AND gate to be counted by the counter giving a reading of 1.0 kHz.

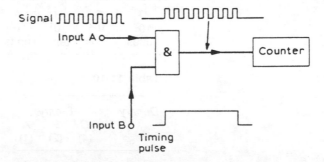

Fig. 11.16 AND gate as frequency meter

Digital clock

A crystal oscillator with very high stability is used to feed a chain of frequency dividers to produce accurate timing pulses. The decoders convert the timing pulses from the dividers into suitable signals to operate the display units (see also Chapter 35). The seconds display is fed with the highest frequency and the hours display with the lowest frequency.

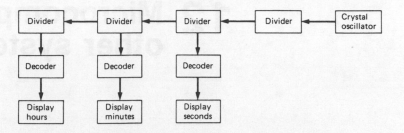

Fig. 11.17 Digital clock system

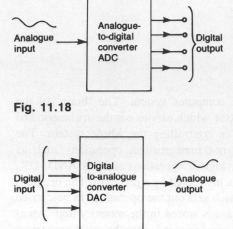

Fig. 11.18

Fig. 11.19

The analogue-to-digital converter

The analogue-to-digital converter (ADC) converts an analogue input into a digitally coded output (Fig. 11.18). The output is a number of parallel digital bits (four in Fig. 11.18), with each bit representing a binary column.

The digital-to-analogue converter

The digital-to-analogue converter (DAC) receives a parallel digital input and converts it back to a voltage (or a current) value that is represented by the binary input. If this is repeated for successive digital inputs, an analogue waveform may be produced (Fig. 11.19).

Digital processing

Most everyday signals such as sound and video are analogue in form. Before such signals are fed into a digital system, e.g. a digital audio tape-recorder (DAT), they have to be converted into a digital form using an analogue-to-digital converter (Fig. 11.20). After processing the digital signal, it is converted back into its original analogue form, e.g. sound signals, using a digital-to-analogue converter.

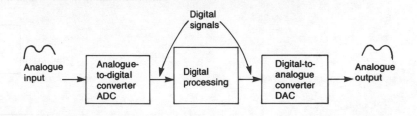

Fig. 11.20

12 Microcomputers and other systems

Computer system

Figure 12.1 shows a basic computer system. The 'brain' of the computer is the microprocessor which carries out the arithmetic and logical operations as well as controlling the whole system. The microprocessor may carry out mathematical operations such as addition and multiplication, logical operations such as AND, OR, NOR etc., or data transfer as instructed by a program. The program is a series of instructions which sets out the operations to be carried out step-by-step. The program is stored in the memory unit. Apart from the program which the memory feeds into the microprocessor, the microprocessor may store some data in memory for later use, e.g. temperature readings at regular intervals for later analysis, or sales figures of a certain item for later comparison with other items. Communication between the microprocessor and the memory, therefore, may take place in both directions, as shown. Such communication is known as *bidirectional*.

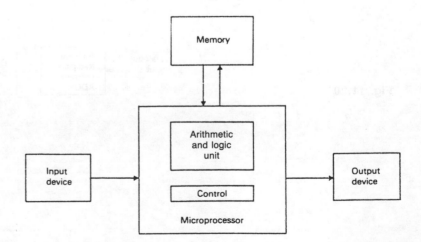

Fig. 12.1

The microprocessor also controls the whole system in terms of timing each step, synchronising the operation of each unit, activating the necessary units and directing the data around the system.

The microprocessor receives input data such as temperature readings, sales or other relevant information as required by the particular program via input devices such as a temperature sensing device or a keyboard. It communicates the results of the program to the outside world via output devices such as a printer or a visual display unit (VDU).

Interface unit

In most applications, direct connection between one part of a system and another is not possible. The signals may be incompatible, requiring conversion (e.g. analogue-to-digital) or conditioning (e.g. attenuation or amplification). To overcome this, an interface is employed between one unit and another, as shown in Fig. 12.2. The interface may be an analogue-to-digital converter, a transducer, an attenuator, or a simple adapter.

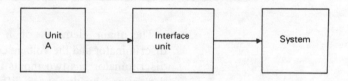

Fig. 12.2

Figure 12.3 shows a complete microprocessor-based computer, also known as microcomputer, including the necessary interface units. Communication to and from the microprocessor is in the form of multi-line parallel connectors carrying digital information in the form of binary numbers. The numbers of parallel lines may be 8, 16, or 32. These parallel connectors are grouped together to form a multi-line **bus**. In most microcomputer systems, peripheral (external) memory stores are used to feed a program or part of a program (sub-routine)

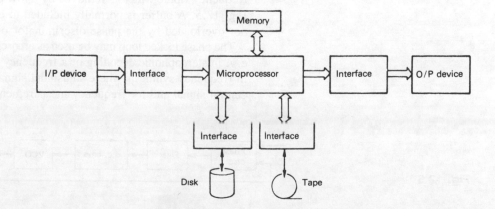

Fig. 12.3

into the microprocessor, to supply information (data), or to store information from the microprocessor. Hence, the bidirectional communication. Such peripheral devices include a tape recorder and a disk which may be used as memory stores to supplement the memory unit of the system.

The phase locked loop

In many applications, phase locking of signals is essential. Phase locking is the process of producing a signal which is in phase with and hence has the same frequency as another reference or control signal. The phase locked loop (Fig. 12.4) is such a phase locking device.

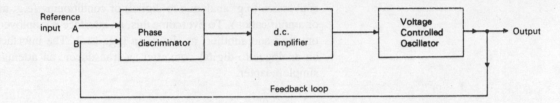

Fig. 12.4

The main elements of a phase locked loop are the phase discriminator and the voltage controlled oscillator (VCO). The phase discriminator has two inputs and a d.c. output. The d.c. output is determined by the phase difference of the two inputs. The VCO produces a sinusoidal output whose frequency is determined by the level of the d.c. input going into it.

Referring to Fig. 12.4, the output from the VCO is fed back to the phase discriminator. A difference in phase between the fed-back signal B and the reference input A produces a d.c. voltage which is amplified before going into the VCO to adjust its frequency. This continues until the fed-back signal B and the reference input A are in phase.

The output from the phase discriminator contains some high frequency ripple which is removed by a low pass filter, as shown in Fig. 12.5. A buffer is normally included to ensure that the VCO is not overloaded by the phase discriminator or any output load.

The phase locked loop may be used as an oscillator with fixed phase, e.g. in stereophonic decoding or a frequency divider. It is also used in frequency synthesis where sinusoidal signals are made up from a central signal and for frequency measurement in control processes.

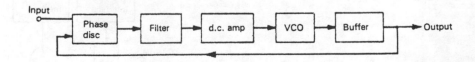

Fig. 12.5

Speed control system

Figure 12.6 shows a block diagram for a speed control system. The purpose of the system is to set the speed of the d.c. motor and keep it constant at that speed. The motor speed is set by the block labelled 'set speed control'. The motor is driven by the d.c. power amplifier.

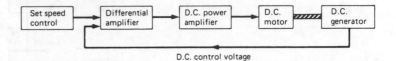

Fig. 12.6

The motor itself drives a d.c. generator, the output of which is determined by the speed of the motor. The output of the d.c. generator is fed back into a differential or difference amplifier. The difference amplifier has two inputs going into it. One input comes from the speed control and the other from the d.c. generator. The difference amplifier compares the two inputs and produces an output which is proportional to the difference between the two inputs. If the motor is running fast, the d.c. control voltage from the d.c. generator is high. The difference between this voltage and the other input to the difference amplifier is therefore reduced. A smaller voltage is therefore obtained from the difference amplifier, reducing the speed of the motor back to its original level. A slow-running motor will produce the opposite effect.

Magnetic tape recorder

In the tape recorder, audio information is converted into a changing magnetic field across a very small gap in a recording head. The changing magnetic field is transformed into a magnetic pattern on a magnetic tape passing along the gap at constant speed. When the tape is played back, the magnetic pattern on the tape is transformed back into a changing magnetic field and subsequently into an audio signal. The recording head itself may be used for playback.

A straightforward magnetisation of the tape introduces non-linearity. To overcome this the audio signal is superimposed on a constant high-frequency signal of between 30 to 100 kHz, referred to as h.f. bias. This high-frequency bias is also used to erase the recording. An erasing head is placed ahead of the recording head which is switched on only when the machine is on record.

Figure 12.7 shows a block diagram of a tape recorder. When switched to playback (a transducer in this case), the playback head converts the magnetic information on the tape into an audio signal. The audio signal is amplified by a pre-amp and fed into a power output stage before going into the loudspeaker. On record, both the record and the erase heads are activated. The audio information from the microphone is amplified before going into a transducer (the record

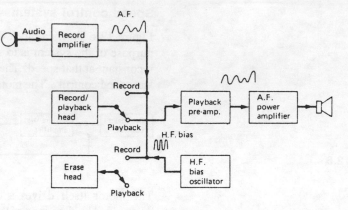

Fig. 12.7

head) to convert it into a changing magnetic field. The erase head ensures that the tape is clean of any magnetic information (before new signals are recorded).

13 Communication

The following four sections deal with the various methods of communication, in other words systems of information transmission. Transmission of information between persons can take many forms. Direct speech is one method — using sound waves through the air. Flag, lamp, or smoke signals are visual methods of communication that normally have a longer range than sound waves. In all these cases, information is conveyed only when a change occurs in the transmitted quantity. A continuous light from a bulb for instance conveys no information. It is only when the light is interrupted according to a certain code that information is passed on.

These systems of communication are limited in the distance they can cover and are dependent on atmospheric conditions. Sound waves can travel a few hundred metres but will be affected by the direction and speed of the wind. Lamp signals depend on clear weather, and so on.

Electrical communication

The use of electricity as a means of communication overcomes many of these difficulties. Electrical communication uses electric waves of one kind or another to transmit the information from one point to another by the use of a wire or without the use of a wire, i.e. wireless transmission.

Transducers

In any electrical communication system, the information to be transmitted such as speech, music or computer data must first be transformed into a changing current or voltage. At the receiving end, the electric wave of current or voltage must be converted back into the original information, i.e. speech, music or coded data. For instance, a microphone is used to transform sound waves into electrical waves, while at the receiving end a loudspeaker is used to convert the electrical waves back into sound waves. Such devices as microphones and loudspeakers are known as transducers.

As with the non-electrical systems of communication mentioned above, a stream of information cannot be transmitted by steady current (or voltage) or by a continuous tone of constant amplitude and frequency. Information can only flow between two points if a steady current or tone is varied in one way or another. The way this is achieved varies with the different methods of electrical communication.

Bandwidth of transmission

The simplest system of electrical transmission is shown in Fig. 13.1.

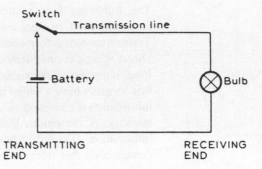

Fig. 13.1 Simple transmission system

It consists of a battery and a key or switch at the sending end. At the receiving end an electrical bulb is used which is connected to the transmitting end by a wire called the transmission line. The simplest code for this system is as follows:

Switch ON   (light on)   means YES.
Switch OFF   (light off)   means NO.

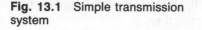

If this information is to be transmitted once every one second, the bulb will go ON and OFF once every second. The waveform transmitted along the line is as shown in Fig. 13.2(a). The frequency of transmission is therefore 1 Hz.

If the information is now increased to 10 (ON and OFF or YES and NO) every second, the waveform will be as shown in Fig. 13.2(b) with 10 complete cycles in a period of 1 second. The frequency is therefore 10 Hz.

If on the other hand the information is to be varied between 1 and 10 (YES and NO) a second, the frequency of transmission will vary between 1 Hz and 10 Hz. In order to accommodate this type of transmission, the system must have a frequency range extending from 1 Hz to 10 Hz. This is known as the bandwidth of the system.

The bandwidth thus depends on the quantity and speed of information. Wider bandwidths are needed for greater amounts of information. Conversely, a system with a given bandwidth can only handle an amount of information determined by the bandwidth itself.

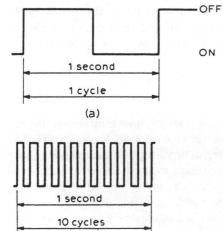

Fig. 13.2

In our previous example, if the bandwidth was limited to 10 Hz, the amount of information that can be transmitted will thus be limited to 10 YESES and NOES.

In practice the bandwidth is made as narrow as possible without damaging the information, in order to reduce costs. For example, a channel having a bandwidth of 300 Hz will occupy as much frequency space in a system as 3 channels each having a bandwidth of 100 Hz. The system with the narrower bandwidths is therefore more economical.

Telephony

The telephone system uses the principle of breaking a steady voltage by an a.f. signal. Figure 13.3 shows a simple block diagram for a telephone system.

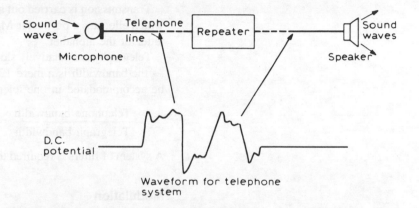

Fig. 13.3 Telephony

Sound waves are transformed into electric waves by the microphone. The telephone line is kept at a d.c. potential of 50 V. This d.c. potential is broken by the a.c. waveform produced by the microphone as shown. An amplifier known as a repeater is used to make good any loss sustained by the signal due to the resistance of the line. More than one repeater may be used depending on the length of the line. At the receiving end, sound waves are reproduced by the speaker.

The telephone system has a bandwidth of 300 Hz to 3400 Hz or 3.1 kHz. This is a small portion of the complete a.f. range of 20 Hz–20 kHz, which is one reason for the poor quality of sound produced by telephones — but it is quite adequate for the purpose of speech.

Telegraphy

Figure 13.4 shows a simplified block diagram for a telegraph system. As the key at the transmitter is closed and opened, a train of pulses is transmitted along the line operating the relay at the receiving end.

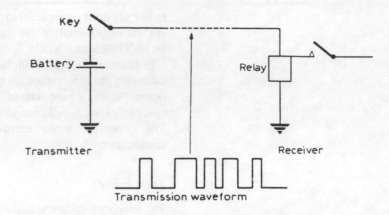

Fig. 13.4 Telegraphy

The relay can be used to operate a switch or a bulb. The waveform of transmission is a sequence of pulses with varying Mark and Space.

Transmission is carried out according to a code (e.g. Morse Code) which allocates a particular Mark and Space time sequence for each letter of the alphabet.

Telegraphy is a relatively slow and cheap system of communication — the bandwidth is a mere 120 Hz, and 25 telegraph channels can be accommodated in one telephone channel:

$$\frac{\text{Telephone bandwidth}}{\text{Telegraph bandwidth}} = \frac{3000}{120} = 25.$$

A system of filters is required to separate one channel from the others.

Modulation

Information can be transmitted from point A to point B by varying an a.c. waveform. The information referred to as the signal is used to change some property of a high frequency wave in such a way as to contain the signal within it. This process is known as modulation.

The high frequency wave is called the **carrier** and after modulation is referred to as the *modulated carrier*.

With modulation, the signal or information is no longer transmitted directly, rather it is 'carried' by the carrier.

Demodulation

At the receiving end, the information has to be extracted from the modulated carrier to reproduce the original signal. This is known as demodulation or detection.

Types of modulation

There are many types of modulation depending on what property of the high frequency carrier is changed by the signal. If the amplitude of the carrier is changed according to the signal, it is known as amplitude modulation. If the frequency of the carrier is changed, it is known as frequency modulation, and so on. The various types of modulation will be dealt with later.

Radio transmission

Radio transmission is wireless transmission using electromagnetic waves, which travel through space at 3×10^8 m/s, the velocity of light.

Radio waves are divided into a number of bands as shown in the table and Fig. 13.5.

VLF (very low frequency)		below 30 kHz
LF (low frequency)	LW (long wave)	30–300 kHz
MF (medium frequency)	MW (medium wave)	300–3000 kHz
HF (high frequency)	SW (short wave)	3000–30 000 kHz
VHF (very high frequency)		30–300 MHz
UHF (ultra high frequency)		300–3000 MHz
SHF (super high frequency)		3000–30 000 MHz
EHF (extra high frequency)		30 000–300 000 MHz

Visible light occupies a small band of the electromagnetic frequency scale (or spectrum) to which the eye is sensitive. Just as the ear is sensitive to a band of sound waves, so the eye is sensitive to a band of electromagnetic waves called light waves. The various sensations of colour are produced by the different frequencies within the light wave band.

Figure 13.5 shows the frequency scale for electromagnetic waves. Up to frequencies of 300 000 MHz, the electromagnetic waves are known as radio waves, which are used as carriers for radio transmission. Radio or wireless transmission uses an aerial to radiate the modulated radio waves and another to receive them.

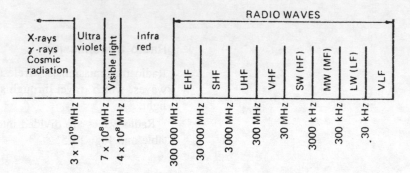

Fig. 13.5 Electromagnetic waves, frequency scale

14 Amplitude modulation

Amplitude modulation (a.m.) is the most commonly used type of modulation. In the a.m. system, the amplitude of the carrier is varied in accordance with the signal or information (Fig. 14.1). With no signal, the carrier has a constant level as in Fig. 14.1(b). When modulated by a sine wave signal, the amplitude of the carrier changes above and below its unmodulated level as the signal goes up and down. The larger the signal the greater is the change in the amplitude of the carrier. The amplitude modulated carrier (Fig. 14.1(c)) has an envelope which is a replica of the modulating signal. When demodulating an a.m. carrier, it is this envelope that is extracted to reproduce the signal.

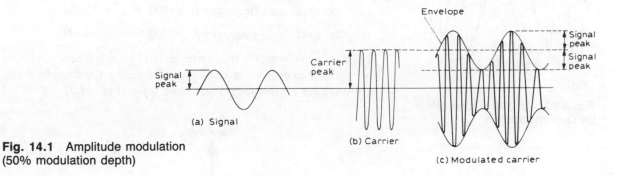

Fig. 14.1 Amplitude modulation (50% modulation depth)

Modulation depth

The ratio of the signal peak to the carrier peak is known as the modulation depth. It tells us the amount by which the level of carrier is changing due to the signal. Modulation depth is always given as a percentage and for this reason it is also referred to as percentage modulation.

$$\text{Modulation depth} = \frac{\text{Signal peak}}{\text{Carrier peak}} \times 100\% \text{ (see Fig. 14.1)}$$

For instance if the signal peak is 1 V and the carrier peak is 2 V,

77

then modulation depth = (1 V)/(2 V) × 100% = 50%.

The a.m. carrier shown in Fig. 14.1 has a modulation depth of 50%.

Over-modulation

Figure 14.2(a) shows an a.m. carrier with 100% modulation. Modulation depths above 100% produce distortion (Fig. 14.2(b)). For this reason modulation depth is limited to a maximum value which for BBC transmission is 80%.

Side frequencies

The amplitude-modulated carrier waveform can be shown to contain three pure sine wave components each of constant amplitude but of a different frequency. The three components are the carrier itself together with the two side frequencies f_1 and f_2. Each signal frequency produces two separate side frequencies. If the signal frequency is f_s and the carrier frequency is f_c, then

$f_1 = f_c - f_s$ known as the lower side frequency

$f_2 = f_c + f_s$ known as the upper side frequency.

For example, if the frequency of the carrier is 100 kHz and that of the signal is 1 kHz, then

the upper side frequency $f_2 = 100 + 1 = 101$ kHz

the lower side frequency $f_1 = 100 - 1 = 99$ kHz.

The a.m. waveform, i.e. the carrier and its two side frequencies, can be represented on what is known as a frequency spectrum by three arrows, each representing a pure sine wave (Fig. 14.3).

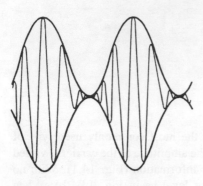

(a)

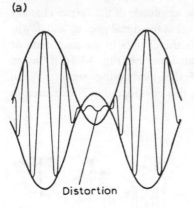

Distortion

(b)

Fig. 14.2 (a) 100% modulation (b) Over-modulation

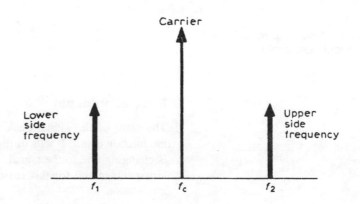

Fig. 14.3 Frequency spectrum for a.m. carrier

Sidebands

Information signals are nearly always of complex form containing a number of pure sine waves. Since each pure sine wave produces

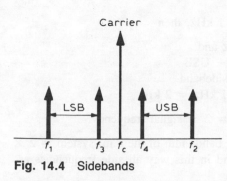

Fig. 14.4 Sidebands

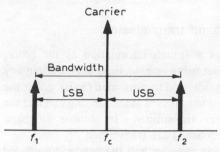

Fig. 14.5 Sidebands and bandwidth for carrier modulated by a single sine wave

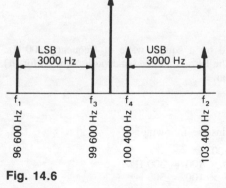

Fig. 14.6

a pair of side frequencies, a complex wave will produce numerous side frequencies and thus create a band of frequencies on either side of the carrier (Fig. 14.4). These are known as the sidebands. The range between the highest and lowest upper side frequencies f_2 and f_4 is known as the upper sideband (USB). Similarly, the range between the highest and lowest lower side frequencies f_3 and f_1 is known as the lower sideband (LSB).

The two sidebands are symmetrical about the carrier and each contain the same information as the other. The carrier contains no information. All the information is in fact carried by the side frequencies.

For a single sine wave signal, the upper and lower sidebands are assumed to extend from the carrier to the upper and lower side frequencies (Fig. 14.5).

EXAMPLE 1

A 100 kHz carrier is amplitude modulated by a signal having a frequency range of 400–3400 Hz. State the sidebands produced.

Solution

The highest signal frequency, 3400 Hz, produces two side frequencies:

$$f_1 = 100\ 000 - 3400 = 96\ 600\ \text{Hz} \quad \text{and}$$
$$f_2 = 100\ 000 + 3400 = 103\ 400\ \text{Hz}$$

(Fig. 14.6).

The lowest signal frequency, 400 Hz, produces a further two side frequencies:

$$f_3 = 100\ 000 - 400 = 99\ 600\ \text{Hz} \quad \text{and}$$
$$f_4 = 100\ 000 + 400 = 100\ 400\ \text{Hz}.$$

$$\text{The upper sideband USB} = f_2 - f_4 = 103\ 400 - 100\ 400$$
$$= 3000\ \text{Hz}$$
$$\text{The lower sideband LSB} = f_3 - f_1 = 99\ 600 - 96\ 600$$
$$= 3000\ \text{Hz}$$

In other words,

$$\text{USB} = \text{LSB} = \text{highest signal frequency} - \text{lowest signal frequency}$$
$$= 3400 - 400 = 3000\ \text{Hz}$$

The side frequencies of any other signal frequency will fall within the upper and lower sidebands.

Bandwidth

Since the side frequencies carry the information, for faithful transmission the bandwidth of the system must be large enough to accommodate all side frequencies present. For a pure sine wave signal, two side frequencies are produced. The bandwidth thus extends from the lower side frequency f_1 in Fig. 14.5 to the upper side frequency f_2.

If the signal frequency is 1 kHz, then

$$
\begin{aligned}
\text{USB} = \text{LSB} &= 1 \text{ kHz and} \\
\text{Bandwidth} &= \text{LSB} + \text{USB} \\
&= 2 \times \text{sideband} \\
&= 2 \times 1 \text{ kHz} = 2 \text{ kHz}
\end{aligned}
$$

In other words, bandwidth = 2 × signal frequency.

For a complex signal, the bandwidth of the a.m. system = 2 × highest signal frequency, and in this way all side frequencies are included.

Single and double sideband transmission

Since one sideband contains as much information as the other, transmission *can* be carried out using one sideband only without any loss of information. In single sideband transmission (SSB), either the lower or the upper sideband is suppressed at the sending end and the only remaining sideband is transmitted. In double sideband transmission (DSB), both sidebands are transmitted.

Single sideband transmission requires half the bandwidth needed for double sideband transmission. For this reason SSB transmission is used in telephony and radio. Twice as many channels can thus be fitted within one band of carrier frequencies. For simplicity DSB transmission is used for all a.m. sound broadcast radio. Normally, DSB transmission is assumed unless otherwise stated.

EXAMPLE

An a.m. carrier is modulated by a square wave of frequency 100 Hz. Neglecting harmonics above the fifth, state the bandwidth required for (*a*) DSB and (*b*) SSB transmission.

Solution

A 100 Hz square wave contains the following:

$$
\begin{aligned}
\text{the fundamental} &= 100 \text{ Hz} \\
\text{the 3rd harmonic} &= 3 \times 100 = 300 \text{ Hz} \\
\text{the 5th harmonic} &= 5 \times 100 = 500 \text{ Hz}
\end{aligned}
$$

Higher harmonics are neglected.
Thus the highest signal frequency = 500 Hz

$$
\begin{aligned}
\text{Bandwidth for DSB} &= 2 \times \text{highest signal frequency} \\
&= 2 \times 500 = 1000 \text{ Hz} \\
\text{Bandwidth for SSB transmission} &= \tfrac{1}{2} \text{DSB} = \tfrac{1}{2} \times 1000 \\
&= 500 \text{ Hz.}
\end{aligned}
$$

The amplitude modulated radio transmitter

In the radio transmitter, the sound waves are first transformed into electric waves (the a.f. modulating signal) by the microphone (Fig.

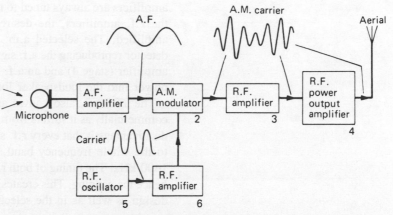

Fig. 14.7 Block diagram of an a.m. transmitter

14.7). The a.f. signal is amplified by an a.f. amplifier (stage 1) to be fed into the a.m. modulator (stage 2). The carrier wave is generated by an r.f. oscillator (stage 5), amplified by an r.f. amplifier (stage 6) tuned to the particular carrier frequency, and fed into the a.m. modulator. The a.m. carrier is then amplified by stage 3 and fed into the r.f. power output amplifier. The output stage drives the r.f. (i.e. carrier) power into the aerial. The bandwidth of the a.m. radio transmission is restricted to 9 kHz using double sideband transmission.

The amplitude modulated radio receiver

A radio receiver is required to operate over a range of carrier frequencies (e.g. the MW and LW bands) and must produce an a.f. output which is a faithful reproduction of the original modulating signal. The receiver must be capable of selecting the desired carrier frequency (i.e. the desired station) to the exclusion of all others.

The tuned radio frequency receiver (TRF)

Figure 14.8 shows a block diagram for a TRF radio receiver. Stages 1 and 2 are two similar amplifiers ganged together. This way both

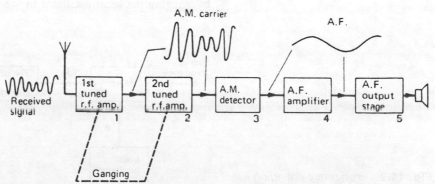

Fig. 14.8 TRF radio receiver

amplifiers are always tuned to the same carrier frequency. By tuning the r.f. amplifiers, the desired carrier frequency is selected and amplified. The selected a.m. carrier is then demodulated by the detector reproducing the a.f. signal. This is followed by an a.f. voltage amplifier (stage 4) and an a.f. output power stage which drives a.f. power into the loudspeaker load.

The TRF set, also known as the straight set, is no longer made commercially as it suffers from a number of disadvantages. The main disadvantage is that every r.f. stage (stages 1 and 2) has to be tunable to the whole frequency band, e.g. for the MW band from 600 to 1600 kHz. The tuning of both r.f. stages has to occur simultaneously, hence the ganging. This creates a number of difficulties in mechanical design as well as in the selectivity of the amplifiers.

The superhet receiver

In order to overcome the disadvantages of the TRF set, the superheterodyne principle was introduced. The carrier frequency selected by the tuner is changed, using a frequency changer, to a common frequency referred to as the **intermediate frequency, the i.f.** The tuned r.f. amplifiers can thus be tuned to one frequency only, namely the intermediate frequency. For a.m. radio sets, an i.f. of 470 kHz is commonly used.

Frequency changing

Frequency changing is achieved by a mixer and a local oscillator (Fig. 14.9). Two separate r.f. signals are fed into the mixer, the received carrier frequency f_c and the local oscillator frequency f_o. The mixer produces four different frequencies: the two original ones, f_c and f_o, together with their sum, $f_o + f_c$, and their difference, $f_o - f_c$. The output is tuned to the difference $f_o - f_c$ and this is the intermediate frequency. The frequency of the local oscillator is made greater than the carrier frequency by the i.f. 470 kHz at all times. This is achieved by ganging the local oscillator to the r.f. tuner.

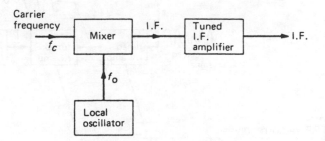

Fig. 14.9 Frequency changing

Block diagram of a superhet receiver

The carrier frequency is selected by the r.f. tuner and subsequently changed to the intermediate frequency by the mixer (Fig. 14.10). Stages 4 and 5 are r.f. amplifiers (known as i.f. amplifiers) tuned to the i.f., 470 kHz. The modulated i.f. is fed into the a.m. detector which extracts the a.f. information feeding it into the a.f. section and so to the loudspeakers.

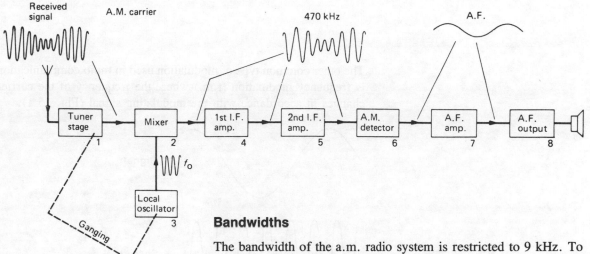

Fig. 14.10 The superhet radio receiver

Bandwidths

The bandwidth of the a.m. radio system is restricted to 9 kHz. To avoid overlap between adjacent transmissions, it follows that each sideband should be only 4.5 kHz, giving a highest signal frequency of 4.5 kHz. The bandwidth of the r.f. tuning stage and the i.f. amplifiers must therefore be 9 kHz to accommodate both sidebands, while that for the a.f. amplifiers need not be higher than 4.5 kHz to accommodate the highest signal frequency.

15 Frequency modulation

The other common type of modulation used in radio communication is frequency modulation (f.m.) where the frequency of the carrier changes in accordance with the modulating signal (Fig. 15.1).

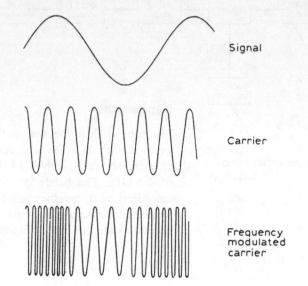

Signal

Carrier

Frequency modulated carrier

Fig. 15.1 Frequency modulation. Note the amplitude of the carrier remaining constant as its frequency changes

Frequency deviation

Frequency deviation is the amount by which the carrier frequency changes for a change of 1 V in the signal level. The unit for frequency deviation is kHz/V. For example, suppose a carrier of frequency 1000 kHz is to be frequency modulated by a square wave having an amplitude of 5 V (Fig. 15.2). If the frequency deviation is given as 10 kHz/V, then for the period from A to B, the carrier frequency will be increased by

$$\text{Signal amplitude} \times \text{Frequency deviation} = 5 \times 10$$
$$= 50 \text{ kHz}$$

giving a carrier frequency of 1000 kHz + 50 kHz
$$= 1050 \text{ kHz}.$$

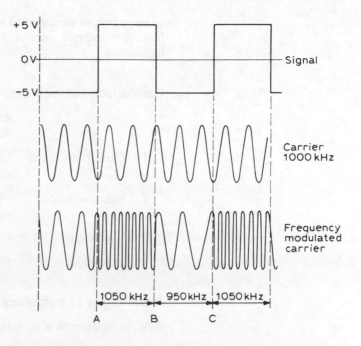

Fig. 15.2 F.M. carrier modulated by a square wave

From B to C the carrier frequency will deviate by a similar amount, namely $5 \times 10 = 50$ kHz, but this time in the negative direction giving a smaller carrier frequency of $1000 - 50 = 950$ kHz.

Maximum deviation

The change in the carrier frequency due to the signal level has to be limited to a maximum value which the carrier is not allowed to exceed. This is known as the maximum deviation. For example, BBC f.m. transmission uses a frequency deviation of 15 kHz/V and a maximum deviation of ± 75 kHz. The maximum signal is limited by the maximum permissible deviation.

$$\text{Maximum signal} = \frac{\text{Maximum deviation}}{\text{Frequency deviation}} = \frac{\pm 75}{15} = \pm 5 \text{ V}$$

or in other words, 5 V in the positive direction or the negative direction.

Side frequencies and bandwidth

When a carrier is frequency modulated with one signal frequency, an indefinite number of side frequencies are produced. The amplitude of the side frequencies gets progressively smaller as they get further away from the carrier.

The bandwidth of an f.m. system has then to be of infinite width to accommodate all the side frequencies. In practice, the smaller side frequencies can be neglected without any noticeable distortion. The BBC for instance uses a bandwidth of 250 kHz for its f.m. transmission.

Comparison between a.m. and f.m. systems

		Amplitude modulation	*Frequency modulation*
1.	Carrier amplitude	Changes with the signal	Remains constant
2.	Side frequencies	Two for each signal frequency	Indefinite number
3.	Bandwidth	9 kHz	250 kHz
4.	Waveband	LW, MW, SW	VHF, UHF

Advantages of frequency modulation

F.M. is superior to a.m. transmission in the following ways:

1. A better quality sound is possible with the f.m. system. This is because of the large bandwidth which can accommodate a greater number of harmonics.
2. F.M. transmission has very low noise level. Noise is unwanted signals that appear at the output, usually in the form of a change in the amplitude of the carrier. In the f.m. system, this can easily be removed by clipping both peaks of the carrier wave. The information contained in the changing frequency of the carrier remains unaltered.

The frequency modulated transmitter

The f.m. transmitter (Fig. 15.3) is similar to the a.m. transmitter except for the modulation (stage 3). Frequency modulation is achieved by an r.f. oscillator whose frequency is varied according to the a.f. signal fed into it from the a.f. amplifier (stage 2).

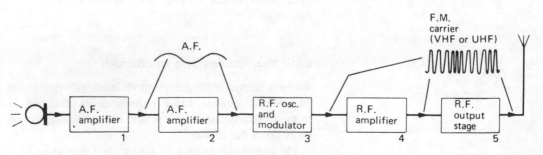

Fig. 15.3   F.M. transmitter

BBC f.m. transmission data

Type	double sideband
Bandwidth	250 kHz
Frequency deviation	15 kHz/V
Maximum deviation	±75 kHz
Carrier	VHF: 89.1, 91.5, 93.5 MHz

The frequency modulated receiver

Figure 15.4 shows a block diagram for a superhet f.m. receiver. The first stage of the f.m. receiver is a tuned r.f. amplifier replacing the tuner in the a.m. receiver. This stage is necessary in f.m. receivers to provide enough amplification of the VHF carrier. The local oscillator together with the mixer (stages 2 and 3) change the selected carrier frequency to an intermediate frequency of 10.7 MHz. This is followed by the i.f. amplifiers. The f.m. detector or decoder (often a ratio detector) translates the frequency deviations into the original a.f. signal which is fed into the a.f. section and so to the loudspeaker load.

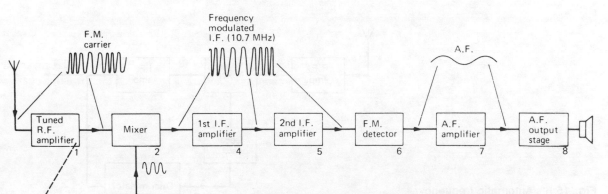

Fig. 15.4 F.M. receiver

The bandwidth of the f.m. system is 250 kHz. This means that the r.f. and the i.f. amplifiers must have the same bandwidth, namely 250 kHz. The a.f. section (stages 7 and 8) has a bandwidth of 20 Hz–20 kHz approximately.

Automatic gain control (a.g.c.)

On many occasions, due to such things as atmospheric conditions, the received carrier level varies or 'fades', causing annoying variations in the output of the receiver. This can be overcome by the use of automatic gain control in both a.m. and f.m. receivers.

A.G.C. is achieved by the d.c. feedback network shown in Fig. 15.5. The d.c. output of the detector is fed back to the input of the first i.f. stage. The feedback voltage controls the gain of the i.f. amplifier in such a way as to increase the gain when the received carrier is weak and decrease it when the received carrier is strong.

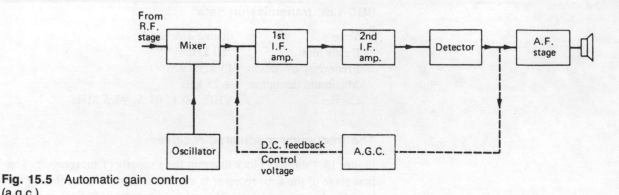

Fig. 15.5 Automatic gain control (a.g.c.)

Automatic frequency control (a.f.c.)

A distorted output from a radio receiver can be caused by incorrect tuning. To obtain the correct tuning, automatic frequency control is used, mainly in f.m. receivers. A.F.C. controls the frequency of the local oscillator, using a discriminator (or comparator) shown in Fig. 15.6. The discriminator translates incorrect intermediate frequencies into a 'control voltage' which adjusts the frequency of the oscillator.

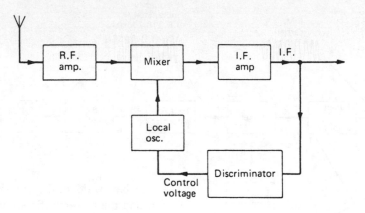

Fig. 15.6 Automatic frequency control (a.f.c.)

16 Pulse modulation

The types of modulation discussed so far have been those employing a continuous carrier. Another type of modulation is possible in which the carrier is discontinuous and exists as a train of pulses. Some characteristic of the pulse is then modified in accordance with the modulating signal.

1. **Pulse amplitude modulation** (PAM) where the amplitude of the pulse changes in accordance with the signal (Fig. 16.1(a)).
2. **Pulse width modulation** (PWM) where the width or duration of the pulse changes in accordance with the signal (Fig. 16.1(b)).
3. **Pulse position modulation** (PPM) where the position of the pulse changes in accordance with the signal (Fig. 16.1(c)).

Pulse code modulation (PCM)

In the pulse amplitude modulation the modulating signal is used to vary the amplitude of a train of pulses. The pulse repetition frequency is constant and so also is the pulse width. Hence it follows that the amplitude variation of the pulses is the only quantity that is significant as far as transmitting the information is concerned. If instead of direct transmission of the pulses, the amplitude of each pulse is indicated

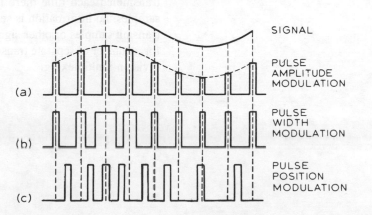

SIGNAL

PULSE AMPLITUDE MODULATION

(a)

PULSE WIDTH MODULATION

(b)

PULSE POSITION MODULATION

(c)

Fig. 16.1 Pulse modulation

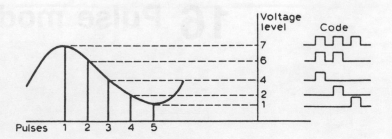

Fig. 16.2 Sampling and coding for PCM system

by some code and this code is then transmitted, the resulting system is known as pulse code modulation (Fig. 16.2).

In this system, the level of each pulse or sample is represented by a series of pulses of constant amplitude and frequency according to a binary code (see Chapter 11). The arrangement of these, i.e. their occurrence or non-occurrence, represents the level of the pulse. For example a 3-digit binary code uses a series of three pulses to represent the level, as shown in Fig. 16.3.

By using a 3-digit binary only 8 discrete levels (i.e. 0—7) can be translated or coded into binary. A 4-digit binary could translate 16 levels (i.e. 0—15) and so on.

At the sending end, an encoding stage is employed to translate each level into a series of pulses. These series of pulses are then transmitted one after the other along the line. At the receiving end, a decoder is used to interpret the coded levels into the original information.

PCM is relatively free from distortion and noise. Its other advantage is that it uses the same coding system as computer data transmission, hence both can be carried out using the same encoding, decoding, amplifying and transmitting equipment.

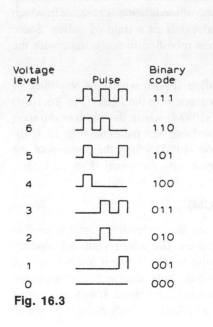

Fig. 16.3

Time division multiplexing

By using an interrupted carrier, i.e. a train of pulses, the modulating signal is sampled at regular intervals and its level is inspected and transmitted each time there is a pulse. In the interval between the samples, no information is sent. These intervals can thus be used to transmit samples of other signals. This way one channel can be used for a number of separate transmissions. This method is known as time division multiplexing.

17 The cathode ray oscilloscope

The cathode ray tube (CRT)

The principle of operation of the CRT is that a negatively charged hot cathode emits electrons which are attracted to and collected by a positively charged anode. This is the principle of the old thermionic valve.

In the CRT, high-speed electrons are emitted by an electron gun (Fig. 17.1). They are focused by an electron lens and directed towards the screen which acts as the positively charged anode. The screen is coated on the inside with fluorescent powder (or phosphor) which gives a visible glow when struck by the high-speed electrons. The electron beam generated by the electron gun produces a stationary spot on the screen. In order to produce a trace, the electron beam has to be deflected in both the horizontal and vertical directions, X and Y.

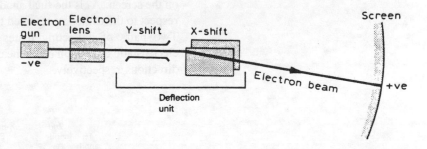

Fig. 17.1 Block diagram for the cathode ray tube

Deflection methods

Two methods are used to deflect the beam in a CRT. The *electrostatic* method using two plates facing each other and an electric potential across them (Fig. 17.2(a)). An electrostatic field is thus created which deflects the electrons passing through it. *Electromagnetic* deflection on the other hand uses the magnetic field created by a current passing through a coil to deflect the electrons. Two sets of deflecting coils

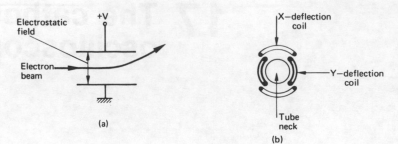

Fig. 17.2 (a) Electrostatic deflection
(b) Electromagnetic deflection

(known as scanning coils in a TV receiver) are used as shown in Fig. 17.2(b). Both methods are capable of producing linear deflection. However, electrostatic deflection has a wider frequency range, hence its use in cathode ray oscilloscopes. Electromagnetic deflection is more suitable for high tube voltages such as used in a TV receiver and requires less space since both coils are placed at the same position along the neck of the tube.

Construction of the CRT

Figure 17.3 shows a schematic diagram of the internal construction of the cathode ray tube using electrostatic deflection. It also shows the various electrodes and their relative potentials. Electrons emitted from the cathode (or electron gun) pass through a small hole (or aperture) in the grid. The grid, which is negative with respect to the cathode potential determines the intensity or number of electrons emitted and thus the brightness of the trace on the screen. The electron beam then passes through an electron lens which focuses the beam on the screen. A_3 is the final anode with a potential of a few kV (with respect to the cathode), referred to as the extra high tension or EHT. Two pairs of deflecting plates D_1 and D_2 are used to provide electrostatic deflection of the beam in the vertical and the horizontal directions respectively.

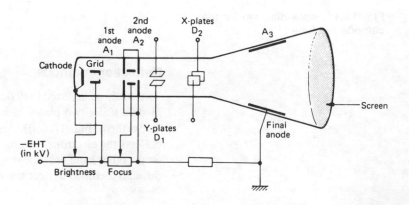

Fig. 17.3 The cathode ray tube

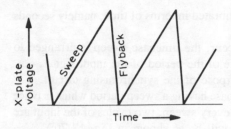

Fig. 17.4 Time base (saw-tooth) waveform

The vertical deflection is produced by the Y-plates, while the horizontal deflection is produced by the X-plates. The input signal is fed to the Y-plates which then deflect the electron beam up and down in accordance with the amplitude of the signal.

The X-plates cause the beam to travel from one side of the screen to the other (the sweep) at a steady pace and then return *very* quickly to its original position (flyback).

The input to the X-shift plates is a saw-tooth waveform shown in Fig. 17.4, generated by an oscillator. This waveform is known as the **time base**.

The appropriate application of signals to both the X and Y shift displaces the electron beam to produce a faithful trace of the input waveform on the screen.

The cathode ray oscilloscope (CRO)

The CRO employs a cathode ray tube together with the necessary devices to produce a stationary trace identical in form to the original input. Figure 17.5 shows a block diagram of the signal and time base circuitry.

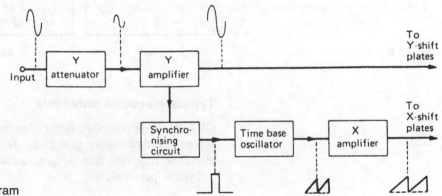

Fig. 17.5 CRO block diagram

The height of the trace is adjusted by the Y-attenuator (or control) and the Y-amplifier giving a calibrated measurement of the amplitude of the input in volts/cm. The time base oscillator is synchronised with the input to the Y-plates to produce a stationary trace on the screen.

Figure 17.6 shows the X- and Y-plates of CRO fed with the signals shown. The signal connected across the Y-plates causes the beam, and therefore the spot on the screen, to move up and down in the vertical direction. The saw-tooth waveform connected across the X-plates causes the beam to move across the screen in the horizontal direction, with a constant speed during the sweep part of the waveform. At the end of the sweep the waveform drops quickly to its starting voltage, returning the spot to its starting point. During each sweep period the horizontal displacement is proportional to time,

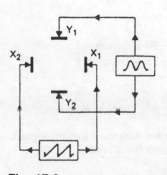

Fig. 17.6

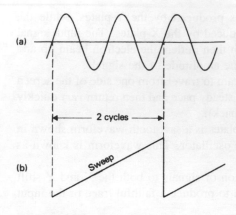

(a)

(b)

2 cycles

Sweep

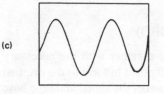

(c)

Fig. 17.7

Fig. 17.8

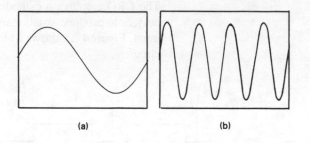

(a)   (b)

thus the timebase may be calibrated in terms of time, namely seconds per centimetre.

For synchronisation to occur, the timebase sweep is arranged to be equal to a whole multiple of the period of the input signal, e.g. 2, 3, 4, etc. This is the purpose of the synchronising circuit.

Figure 7.17 shows a time base having a sweep period which is twice that of the input signal. For every sweep, two cycles of the input are covered; hence the display will be as shown in Fig. 17.7(c).

If the timebase frequency is now doubled, i.e. the sweep is halved, making it equal to the period of the input, the trace will be as shown in Fig. 17.8(a), displaying one cycle only. Conversely, if the timebase frequency is halved (i.e. doubling the sweep duration), the trace will cover twice as many cycles of the input in this case, displaying four cycles, as shown in Fig. 17.8(b).

Typical electrode potentials

CRT electrode potentials differ, depending on the size of the CRT screen. Further higher potentials are necessary in tubes used in television receivers than in those used in oscilloscopes.

Typical potentials are:

	Oscilloscope	*Television*
Cathode	30 V	70 V (mono)
		120 V (colour)
Grid	0 V	30 V
Focusing lens	+1400 V	2−3 kV
Final anode	+2000 V	15−20 kV (mono)
		20−25 kV (colour)

EXAMPLE 1

The X and Y plates of an oscilloscope are fed with the following signals. For each case, sketch the expected display on the screen.

Solution

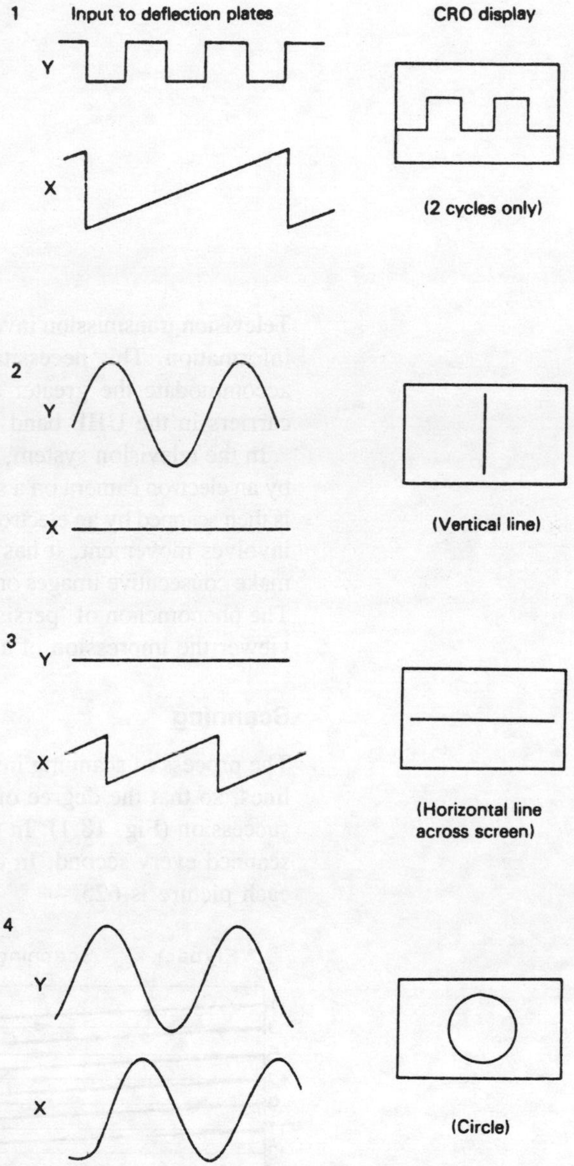

1	Input to deflection plates	CRO display

Y

X

(2 cycles only)

2

Y

X

(Vertical line)

3

Y

X

(Horizontal line across screen)

4

Y

X

(Circle)

Fig. 17.9

18 Television

Television transmission involves transmitting picture as well as sound information. This necessitates a system with a wide bandwidth to accommodate the greater amount of information. For this reason carriers in the UHF band are used.

In the television system, the picture to be transmitted is produced by an electron camera on a specially formed light-sensitive plate which is then scanned by an electron beam. As the transmitted scene normally involves movement, it has to be scanned in a time short enough to make consecutive images only very slightly different from each other. The phenomenon of 'persistence of vision' in the eye then gives the viewer the impression of a moving picture just as a cine film does.

Scanning

The process of scanning involves exploring the picture in a series of lines, so that the degree of brightness is examined at each point in succession (Fig. 18.1). In the TV system, 25 complete pictures are scanned every second. In the British system the number of lines in each picture is 625.

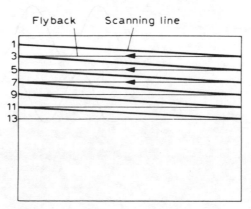

Fig. 18.1 Scanning lines

Interlacing

Although 25 pictures per second produce the effect of continuous motion, such a picture suffers from severe flicker. This can be overcome by increasing the picture frequency to 50 pictures per second. As this would double the bandwidth required, the method of interlacing is used. First, all the odd lines are scanned, followed by the even ones. Each half picture is known as a 'field' or 'frame'. Hence there are 50 fields every second. The eye sees 50 half pictures every second which are nearly identical — the effect is virtually the same as seeing 50 pictures a second, and this removes the flicker.

The video waveform

The video waveform consists of the entire sequence of picture signals as scanned line by line. The start of each line and field is indicated by the use of synchronising pulses (sync pulses). Two different sync pulses are used: a line sync pulse at the beginning of each line (Fig. 18.2) and a field sync pulse at the beginning of each field. The field sync pulses occur 50 times every second giving a field frequency of 50 Hz. The line sync pulses have a much higher repetitive rate. Using the 625 British system, the line sync pulses occur 625 times for every complete picture giving $625 \times 25 = 15\ 625$ lines per second or a line frequency of 15 625 Hz (15.625 kHz).

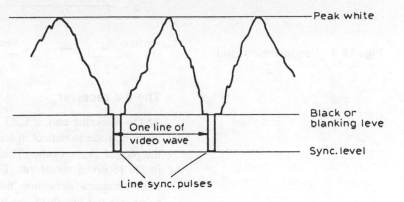

Fig. 18.2 Video waveform

The sync pulses are all below the black level as can be seen in Fig. 18.2. The flyback thus takes place while the tube is blanked out, producing no trace on the screen.

The 625 system

This system uses 625 lines per picture.

Carrier	UHF band (400–900 MHz)
Modulation	video is a.m., sound is f.m.
Video bandwidth	5.5 MHz
Field frequency	50 Hz
Line frequency	15.625 kHz

Vestigial sideband transmission

The method of modulation in use for transmitting TV signals is amplitude modulation. Ordinary a.m. gives rise to two sets of sidebands on either side of the carrier. Since one sideband contains the same information as the other, the information contained in the modulated carrier can be conveyed by one sideband only, thus effectively halving the bandwidth. However, in practice it is impossible to achieve wideband modulation with one sideband completely eliminated. A gradual cut-off of one sideband is employed as shown in Fig. 18.3. The reduction of information by the gradual cut-off of the upper sideband between the carrier frequency f_c and f_2 is made up by transmitting a vestige (i.e. a part) of the lower sideband f_1 to f_c. This is known as vestigial sideband transmission.

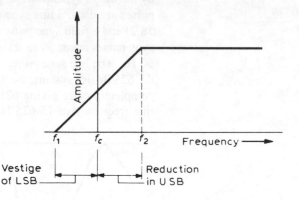

Fig. 18.3 Vestigial sideband

The TV receiver

At the receiving end, a CRT is used to reproduce the picture. The beam is made to deflect in accordance with the original scanning of the picture at the studio. This is secured by the sync pulses present in the received waveform. Unlike the CRT used in oscilloscopes, electromagnetic deflection, using *coils*, is employed for TV receivers to reduce the length of the tube.

Block diagram (Fig. 18.4)

The radio wave signal is received by the aerial and the particular station is selected by the r.f. amplifier (stage 1) which also provides the first stage of amplification of the carrier. The mixer-oscillator (stages 2 and 3) changes the carrier frequency to an intermediate frequency of 39.5 MHz. The i.f. is amplified by the i.f. amplifier (stage 4) and fed into the a.m. detector (stage 5) for demodulation. The detected signal contains the video or picture information together with the sync pulses and the sound information. These three component parts have to be separated and fed into their individual channels.

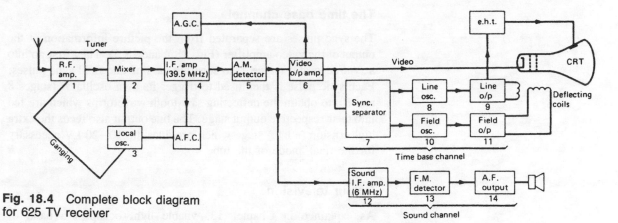

Fig. 18.4 Complete block diagram for 625 TV receiver

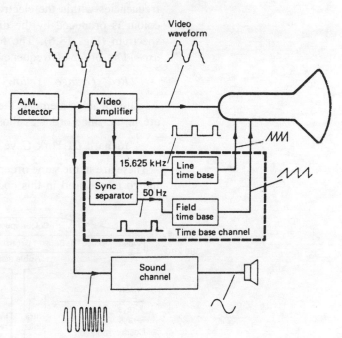

Fig. 18.5 Simplified block diagram for 625 TV receiver

At the output of the a.m. detector, the f.m. sound is separated and fed into the sound channel (see also Fig. 18.5). The remaining video and sync pulses are amplified by the video amplifier to be separated at its output. The picture information goes directly to the cathode of the CRT, while the sync pulses are fed into the time base channel.

The sound channel

The sound signal is in the form of a frequency modulated carrier at an intermediate frequency of 6 MHz. This is amplified (stage 12), detected by an f.m. detector (stage 13), and passed through the a.f. output stage (stage 14) to the loudspeaker.

The time base channel

The sync pulses are separated from the picture information at the output of the video amplifier (Figs 18.4 and 18.5). They first go into a sync separator (stage 7) to separate the line and field sync pulses. Each sync pulse is then used to trigger its own oscillator (stages 8 and 10) to obtain the deflecting saw-tooth waveforms which are fed into their respective output stage. The line output also feeds the extra high tension (e.h.t.) stage which provides the 10−20 kV necessary for the final anode of the tube.

Colour television

As explained in Chapter 13, visible light occupies a band of frequencies within the electromagnetic spectrum. The sensation of colour is produced by the different frequencies within the visible spectrum (Fig. 18.6). The following colours can be observed, in *ascending* order of frequency:

Red   Orange   Yellow   Green   Blue   Indigo   Violet

As a useful aid to memory, the initial letters of the following sentence are in the same order as those for the spectrum colours:

*R*ichard *O*f *Y*ork *G*ave *B*attle *I*n *V*ain

These are in the same order as the resistor colour code except that indigo is not used in this code. Red is 2, orange is 3, and so on.

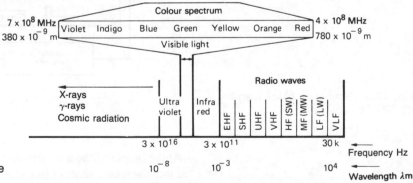

Fig. 18.6 Electromagnetic wave spectrum

Primary colours

Colours can be created by mixing other colours together. In colour TV, the various colours are produced by mixing *red, green* and *blue* (R, G, B) in different proportions. These three colours are known as primary colours. While an infinite variety of colours can be created by different mixtures of R, G and B, the following general rules should be committed to memory:

$R + G = $ *yellow*
$R + B = $ *magenta*
$B + G = $ *cyan*
$R + G + B = $ *white*

Colour transmission

One basic requirement of colour TV transmission is that a monochrome (black and white) TV set should be capable of receiving colour transmission and producing a normal black and white picture without any modification to the set. This is known as compatibility between colour and monochrome transmissions.

In order to meet this requirement, colour TV transmission is carried out by sending the monochrome (known as the luminance) information in the normal way. The colour content (known as the chrominance) is used to modulate a sub-carrier of 4.43 MHz added to the monochrome signal as shown in Fig. 18.7.

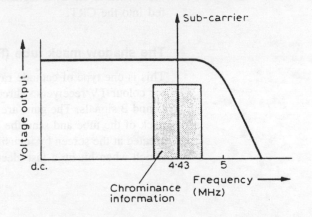

Fig. 18.7 Colour TV transmission. A sub-carrier is used for the colour content

At the studio, three different cameras are used to explore the colour content of the picture. An appropriate filter is placed in front of each allowing only one primary colour through, e.g. a red filter for the 'red' camera and similarly for green and blue.

The TV colour receiver

Figure 18.8 shows a block diagram for a colour TV receiver. For clarity, the time base and sound channels are not shown since they are identical to the monochrome receiver. The modulated carrier is received and detected by the a.m. detector. The video signal reproduced by the a.m. detector contains two components: the monochrome or luminance (Y) and the colour content or chrominance. The luminance is amplified separately and fed into the decoder. The decoder contains a chrominance amplifier, detector and matrix. It

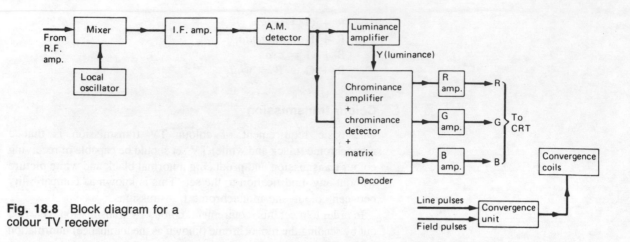

Fig. 18.8 Block diagram for a colour TV receiver

receives both the luminance and chrominance signals and converts them back to R, G, B signals, which are amplified separately and fed into the CRT.

The shadow mask tube (Fig. 18.9)

This is one type of cathode ray tube used by British manufacturers for colour TV receivers. Three identical guns are used, fed with R, G and B signals. The guns are mounted at 120° to each other at the neck of the tube and scan the screen simultaneously. The colour is created at the screen by focusing each gun on a specially coated dot which when hit by a fast electron emits a primary colour.

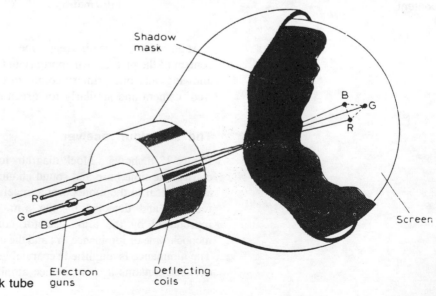

Fig. 18.9 The shadow mask tube

These dots are arranged to form a regular system of triangles known as **triads** (RGB). To ensure that each gun hits its particular dot and no other, a 'shadow mask' is used. The dots are indistinguishable to the naked eye which sees the three colours mixed together, recreating the original coloured picture. Convergence coils fed with line and field pulses placed on the neck of the tube ensure that the three guns scan the same line.

The in-line colour tube

In the in-line shadow-mask tube the three guns are placed side by side and the phosphorous coating on the screen is in the form of striped triads. Each 3-colour triad is arranged to coincide with a longitudinal grill or slot in the shadow mask.

The first in-line tube was developed by Sony, and was known as the **trinitron**. This was followed by Mullard's precision-in-line, PIL (AX series) self-converging tube (Fig. 18.10).

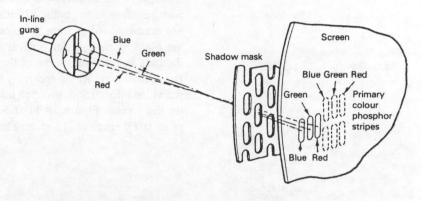

Fig. 18.10   The precision-in-line (PIL) tube

NICAM digital stereo

As stated earlier, sound is transmitted using a 6 MHz frequency modulated carrier which, falling just outside the video bandwidth, prevents any interference. This method can be used to provide a good quality sound, given good quality sound amplification at the receiver. However, it is incapable of producing hi-fi quality and is unable to carry stereophonic sound. It is not possible to add a second sound carrier without causing unacceptable interference to either the vision or the sound carriers. In order to achieve stereophonic, hi-fi sound a new system had to be developed.

Following years of research and development, BBC engineers came out with a radically new sound system for TV broadcasting known as NICAM 728, or NICAM for short. NICAM stands for Near Instantaneous Companded Audio Multiplex and 728 refers to the data rate of 728 kbit/s. It provides two completely independent sound channels so that dual-language sound tracks may be transmitted as well as stereophonic sound. It can carry information in one or both channels and is completely separate and independent of the existing 6 MHz f.m. monophonic sound channel.

High definition television (HDTV)

A high definition television system attempts to produce a television picture to a quality approaching that of a 16 mm film. To do this, a very much higher number of scan lines is necessary. It is proposed that 1250 scan lines are used together with a new elongated (letter box) picture, thus increasing the total bandwidth to 34.7 MHz.

Advanced and complex digital processing may then be used to reduce the actual transmitted bandwidth to a more manageable size. Because of the enlarged bandwidth, satellite transmission has to be used, employing microwave frequencies above 1000 MHz (1 GHz).

Teletext

Teletext is the transmission of coded data in addition to the normal picture information. In the 625 line TV system, only about 575 lines are actually used to form the picture on the screen. The remainder are blanked to allow for vertical scan and synchronisation. Teletext data are inserted into some of these unused lines and transmitted in the normal way. At the receiver, the teletext data are separated from the video information, grouped into 'pages' and processed for display on the screen. Figure 18.11 shows a simplified arrangement for a teletext TV receiver. To receive teletext information, the video signal

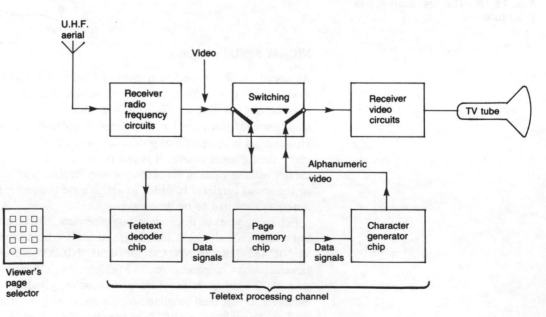

Fig. 18.11

is diverted into the teletext decoder by an electronic switch. The decoder, which consists of a number of silicon chips, first arranges the data into groups of rows and pages in accordance with the user's instruction, stores the information into a memory chip and uses a character generator to display the information in the form of alphanumeric characters on the screen.

Digital TV receiver

Apart from teletext and NICAM, the video information itself may be processed using advanced digital technology. Figure 18.12 shows the essential elements of a 'digital' TV receiver. The composite video from the i.f. stage is fed into an analogue-to-digital converter (ADC) to translate the analogue signal into a series of coded digital pulses. The digital video signal is then used to provide the input to the video processing and deflection processing channels.

The video processor chip carries out all necessary luminance and chrominance processing functions. The output of the video processor is decoded to produce the digitised Red, Green and Blue (RGB) signals which are fed into a digital-to-analogue converter (DAC) to re-create the original analogue RGB signals before going into the colour tube.

The deflection processor chip provides timebase synchronisation and line and field drive waveforms which feed into their respective output stages before going into the scan coils. 'Digital' TV receivers

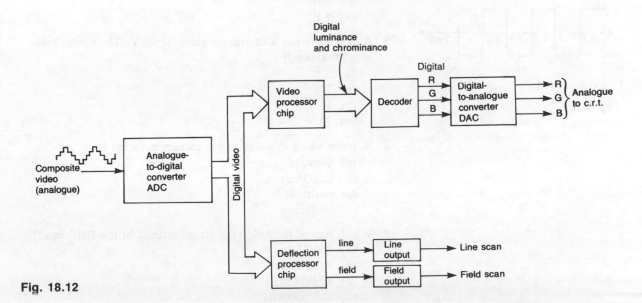

Fig. 18.12

usually employ microcomputer control to ensure very accurate line and field frequencies, resulting in a very steady picture as well as improved quality.

19 Useful Exercises I

CHAPTERS 1–18. For answers see Appendix VII.

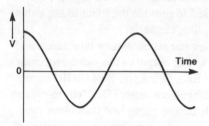

Fig. Q.1

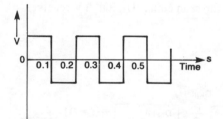

Fig. Q.2

1 The number of cycles shown in the sinusoidal waveform in Fig. Q.1 is
  (*a*) $1\frac{1}{4}$
  (*b*) $1\frac{1}{2}$
  (*c*) $2\frac{1}{2}$
  (*d*) 3

2 The frequency of the waveform shown in Fig. Q.2 is
  (*a*) 2 Hz
  (*b*) 2.5 Hz
  (*c*) 4 Hz
  (*d*) 5 Hz

3 A sine wave has a maximum value of 10 V. The r.m.s. value is approximately
  (*a*) 5 V
  (*b*) 7 V
  (*c*) 10 V
  (*d*) 20 V

4 In pulse width modulation the carrier is varied in
  (*a*) duration
  (*b*) amplitude
  (*c*) position
  (*d*) frequency

5 Which one of the following frequencies is in the UHF band?
  (*a*) 100 MHz
  (*b*) 1000 MHz
  (*c*) 10 000 MHz
  (*d*) 100 000 MHz

6 The BBC f.m. radio transmission uses a maximum deviation of
  (*a*) ±9 kHz
  (*b*) ±30 kHz

(c) ±75 kHz

(d) ±250 kHz

7 On the medium wave transmission, a.m. is used rather than f.m. because it has

(a) reduced bandwidth

(b) better sound quality

(c) less noise

(d) greater volume

8 When a sine wave is applied to a differentiator the output is

(a) small duration pulses

(b) square

(c) triangular

(d) sinusoidal

9 The circuit contained in the block in Fig. Q.3 is

(a) an integrator

(b) a clipper

(c) a d.c. restorer

(d) an interstage coupler

Fig. Q.3

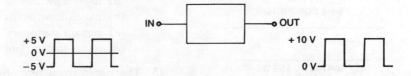

10 Which of the symbols in Fig. Q.4 represents a pnp transistor?

Fig. Q.4

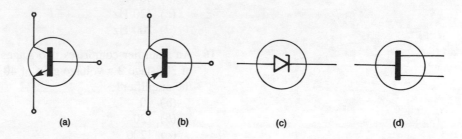

(a)   (b)   (c)   (d)

11 The type of modulation used for the sound channel of the 625-line TV transmission is

(a) a.m.

(b) phase modulation

(c) pulse code modulation

(d) f.m.

12 In Fig. Q.5, block 2 represents

(a) an i.f. amplifier

(b) a local oscillator

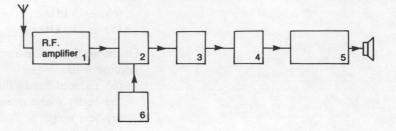

Fig. Q.5

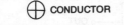

MAGNETIC FIELD

Fig. Q.6

(*c*) a mixer

(*d*) a detector

13 What type of logic element or combination of elements could be used to provide a 1 output only when four 0 inputs are applied?

(*a*) 2 input NOR gate plus a 2 input AND gate

(*b*) 2 input NAND gate plus a 2 input NOR gate

(*c*) 4 input NOR gate

(*d*) 4 input NAND gate

14 Refer to Fig. Q.6. If the direction of the current in the conductor is away from the viewer the conductor will

(*a*) move upwards

(*b*) move downwards

(*c*) remain stationary

(*d*) move to the right

15 The ripple frequency observed on the output of a full wave rectifier is 120 Hz. What is the frequency of the mains input?

(*a*) 50 Hz

(*b*) 60 Hz

(*c*) 120 Hz

(*d*) 200 Hz

16 An amplifier comprises two stages. Stage 1 has a voltage gain of 30, stage 2 a voltage gain of 40. What is the overall gain of the amplifier?

(*a*) 70

(*b*) 120

(*c*) 1200

(*d*) 7000

17 The main purpose of the regulator in Fig. Q.7 is to compensate for

(*a*) transformer losses

(*b*) load variations

(*c*) inefficient rectifier

(*d*) frequency variations

Fig. Q.7

18 The block in Fig. Q.8 represents
   (*a*) a clipper
   (*b*) an a.c. coupler
   (*c*) a d.c. restorer
   (*d*) a rectifier

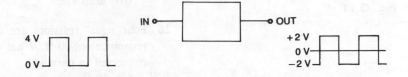

Fig. Q.8

19 The type of feedback used to reduce distortion of the signal at the output of an amplifier is
   (*a*) negative from output to input
   (*b*) positive from output to input
   (*c*) negative from input to output
   (*d*) positive from input to output

20 A circuit which produces an alternating waveform when supplied with d.c. power is called
   (*a*) a modulator
   (*b*) a detector
   (*c*) a rectifier
   (*d*) an oscillator

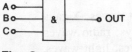

Fig. Q.9

21 The output from the gate shown in Fig. Q.9 will be available when
   (*a*) inputs A and C only are present
   (*b*) inputs B and C only are present
   (*c*) inputs A, B and C are present
   (*d*) any two of the inputs are present

22 A continuity check on a device produces a reading of 500 kΩ. When the device is reversed the reading is 50 Ω. The device is a
   (*a*) thermistor
   (*b*) capacitor
   (*c*) v.d.r.
   (*d*) germanium diode

23 The meters shown in Fig. Q.10 will indicate voltage and current as follows:
   (*a*) M_2 — voltage M_1 — current
   (*b*) M_2 — current M_1 — current
   (*c*) M_2 — current M_1 — voltage
   (*d*) M_2 — voltage M_1 — voltage

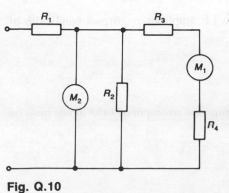

Fig. Q.10

24 The effect of disconnecting the a.g.c. in a radio receiver is that the output
   (*a*) will be of reduced strength
   (*b*) will be loud and distorted
   (*c*) will vary considerably
   (*d*) stage will be damaged

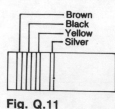

Fig. Q.11

25 The resistor shown in Fig. Q.11 may vary from the coded value by
  (a) ±2 kΩ
  (b) ±5 kΩ
  (c) ±10 kΩ
  (d) ±20 kΩ

26 Four equal resistors are connected in series and their total resistance is 800 Ω. What would be their total resistance when connected in parallel?
  (a) 10 Ω
  (b) 50 Ω
  (c) 400 Ω
  (d) 1600 Ω

27 Two strips, A and B, of identical metal have the same length and thickness. If the width of A is half that of B, the ratio of the resistance of A to the resistance of B will be
  (a) 1 : 2
  (b) 1 : 4
  (c) 2 : 1
  (d) 4 : 1

28 In which one of the following are the signals in ascending frequency?
  (a) Sound waves, radio waves, light waves, X-rays
  (b) Sound waves, light waves, X-rays, radio waves
  (c) X-rays, radio waves, sound waves, light waves
  (d) Light waves, X-rays, sound waves, radio waves

29 If the frequency range of 100−5000 Hz is used to amplitude-modulate a carrier the resulting bandwidth is
  (a) 100 Hz
  (b) 400 Hz
  (c) 5000 Hz
  (d) 10 000 Hz

30 In an f.m. receiver, the i.f. amplifier requires a bandwidth of
  (a) ±9 kHz
  (b) ±100 kHz
  (c) ±10.7 MHz
  (d) ±470 kHz

31 Refer to Fig. Q.12. To produce an output of 1 the inputs must be

	E	F	G	H
(a)	0	0	1	1
(b)	1	1	0	0
(c)	1	1	1	0
(d)	0	0	0	1

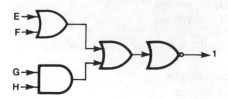

Fig. Q.12

32 In an a.m. system the effect of varying the modulating voltage is to alter the
(*a*) bandwidth
(*b*) depth of modulation
(*c*) side frequencies
(*d*) deviation

33 A 10 Ω resistor dissipates 90 W when connected to a 30 V a.c. supply. The supply voltage is the
(*a*) r.m.s. value
(*b*) average value
(*c*) maximum value
(*d*) peak-to-peak value

34 Which of the following binary codes represents the number 7?
(*a*) 0110
(*b*) 0111
(*c*) 1001
(*d*) 1110

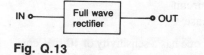

Fig. Q.13

35 Refer to Fig. Q.13. With a sinusoidal input, the waveform at the output is

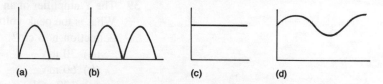

(a)   (b)   (c)   (d)

36 If the amplitude of the modulating signal in an f.m. system is increased, the effect on the carrier would be
(*a*) an increased amplitude
(*b*) an increased depth of modulation
(*c*) an increased deviation
(*d*) a reduction in the mean carrier frequency

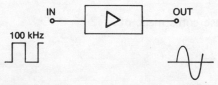

100 kHz

Fig. Q.14

37 Refer to Fig. Q.14. The frequency response of the amplifier is

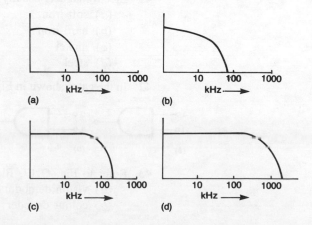

CATHODE RAY OSCILLOSCOPE

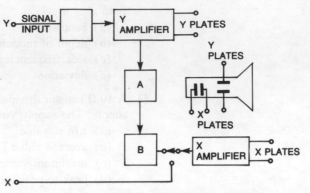

Fig. Q.15

38 Refer to Fig. Q.15 which is a block diagram of a cathode ray oscilloscope. Blocks A and B are

	A	B
(a)	sync circuit	attenuator
(b)	sync circuit	time base
(c)	time base	sync circuit
(d)	attenuator	time base

39 The Y-amplifier of an oscilloscope has a sensitivity of 10 mV/cm. What is the peak voltage of a displayed sinewave if the total Y-deflection is 6 cm?
  (a) 30 mV
  (b) 60 mV
  (c) 120 mV
  (d) 600 mV

40 In an f.m. receiver, a.f.c. may be applied to a
  (a) a.f. output
  (b) f.m. detector
  (c) local oscillator
  (d) r.f. amplifier

41 Electromagnets usually use cores made of
  (a) soft iron
  (b) hard iron
  (c) nickel
  (d) steel

42 In the gates shown in Fig. Q.16, which will have an output of 1?

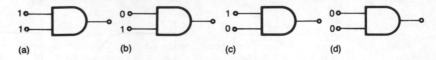

Fig. Q.16   (a)   (b)   (c)   (d)

43 Refer to Fig. Q.17. Block X is a
  (a) vision demodulator
  (b) colour decoder

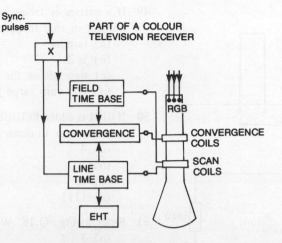

Sync. pulses

PART OF A COLOUR
TELEVISION RECEIVER

X

FIELD
TIME BASE

CONVERGENCE

LINE
TIME BASE

EHT

RGB

CONVERGENCE
COILS

SCAN
COILS

Fig. Q.17

(c) sync separator

(d) chrominance amplifier

44 A square wave contains the following harmonics in addition to the fundamental:

(a) even harmonics only

(b) odd harmonics only

(c) even and odd harmonics

(d) no harmonics

45 The bandwidth of a video amplifier is

(a) 20 Hz−20 kHz

(b) d.c. − 5 MHz

(c) 470 kHz−10.7 MHz

(d) 10 MHz−40 MHz

46 Which of the following materials is a good insulator?

(a) Germanium

(b) Silver

(c) Dry wood

(d) Steel

47 If a capacitor of 1 farad has a charge of 10 coulombs, what is the voltage across it?

(a) 0.1 V

(b) 1 V

(c) 10 V

(d) 100 V

48 If the load current drawn from an unstabilised supply is initially small and then increased, the output voltage will

(a) be unchanged

(b) decrease

(c) increase

(d) increase initially and then decrease

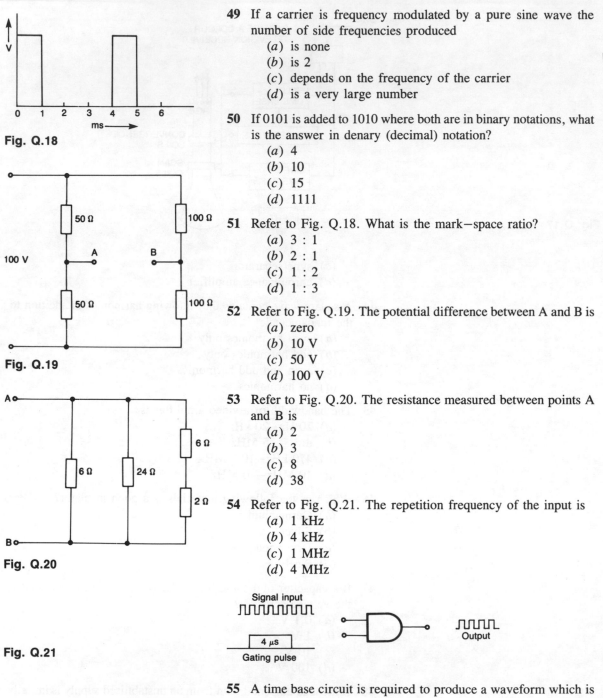

Fig. Q.18

Fig. Q.19

Fig. Q.20

Fig. Q.21

49 If a carrier is frequency modulated by a pure sine wave the number of side frequencies produced
   (a) is none
   (b) is 2
   (c) depends on the frequency of the carrier
   (d) is a very large number

50 If 0101 is added to 1010 where both are in binary notations, what is the answer in denary (decimal) notation?
   (a) 4
   (b) 10
   (c) 15
   (d) 1111

51 Refer to Fig. Q.18. What is the mark–space ratio?
   (a) 3 : 1
   (b) 2 : 1
   (c) 1 : 2
   (d) 1 : 3

52 Refer to Fig. Q.19. The potential difference between A and B is
   (a) zero
   (b) 10 V
   (c) 50 V
   (d) 100 V

53 Refer to Fig. Q.20. The resistance measured between points A and B is
   (a) 2
   (b) 3
   (c) 8
   (d) 38

54 Refer to Fig. Q.21. The repetition frequency of the input is
   (a) 1 kHz
   (b) 4 kHz
   (c) 1 MHz
   (d) 4 MHz

55 A time base circuit is required to produce a waveform which is
   (a) sinusoidal
   (b) d.c.
   (c) saw-tooth
   (d) short duration pulses

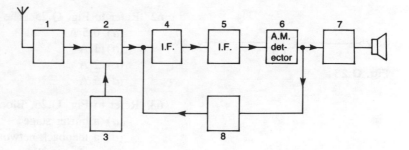

Fig. Q.22

56 Refer to Fig. Q.22. Block 1 represents a
(a) mixer
(b) r.f. tuner
(c) a.f. stage
(d) frequency discriminator

57 Refer to Fig. Q.22. Block 3 represents
(a) a local oscillator
(b) a mixer
(c) an a.f.c.
(d) a signal limiter

58 Refer to Fig. Q.22. Block 8 represents
(a) an a.f. output stage
(b) a sync separator
(c) an a.g.c.
(d) an a.f.c.

59 The symbol shown in Fig. Q.23 represents
(a) an amplifier
(b) an oscillator
(c) a high pass filter
(d) a low pass filter

60 An integrator
(a) is a low pass filter
(b) is a high pass filter
(c) converts d.c. into a.c.
(d) is an f.m. detector

61 Refer to Fig. Q.24. Block A represents
(a) an f.m. demodulator
(b) an a.m. detector
(c) an audio amplifier
(d) a de-emphasis network

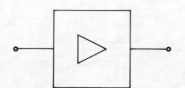

Fig. Q.23

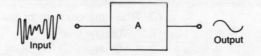

Fig. Q.24

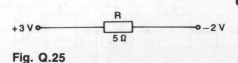

Fig. Q.25

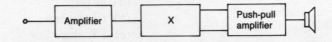

Fig. Q.26

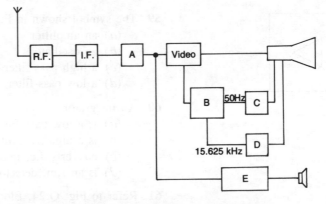

Fig. Q.27

62 Refer to Fig. Q.25. The current through the resistance is
   (*a*) 0.5 A
   (*b*) 1 A
   (*c*) 2 A
   (*d*) 5 A

63 Refer to Fig. Q.26. Block X represents
   (*a*) a buffer stage
   (*b*) a feedback network
   (*c*) a phase splitter
   (*d*) a voltage amplifier

64 Refer to Fig. Q.27. Block A represents
   (*a*) an a.f. amplifier
   (*b*) an f.m. decoder
   (*c*) a pulse generator
   (*d*) an a.m. detector

65 Refer to Fig. Q.27. Block C represents
   (*a*) a decoder
   (*b*) a field time base
   (*c*) a line time base
   (*d*) a sync separator

66 Refer to Fig. Q.27. Block E represents
   (*a*) a sound channel
   (*b*) a time base channel
   (*c*) a negative feedback network
   (*d*) a colour discriminator

67 The time constant of the waveform shown in Fig. Q.28 is
   (*a*) t_1
   (*b*) t_2
   (*c*) t_3
   (*d*) t_4

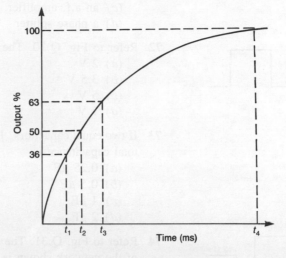

Fig. Q.28

68 The advantage of using the 625 TV system over the 405 system is that it
  (a) simplifies the aerial
  (b) uses cheaper receivers
  (c) produces a better quality picture
  (d) can be used for colour

69 Increased frequency deviation of the input to a discriminator will produce an output having a
  (a) higher audio volume level
  (b) higher audio frequency
  (c) lower audio volume level
  (d) lower audio frequency

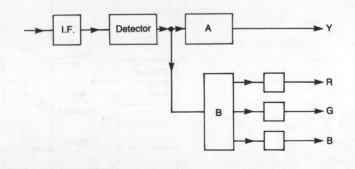

Fig. Q.29

70 Refer to Fig. Q.29. Block A represents
  (a) a time base channel
  (b) a sound channel
  (c) a decoder
  (d) a luminance amplifier

71 Refer to Fig. Q.29. Block B represents
  (a) a matrix decoder
  (b) a sync separator

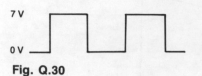

Fig. Q.30

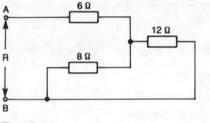

Fig. Q.31

(c) an a.f. amplifier

(d) a phase splitter

72 Refer to Fig. Q.30. The average voltage of the waveform is

(a) 2 V

(b) 3.5 V

(c) 5 V

(d) 7 V

73 If two equal capacitors, 1 μF each, are connected in series, the total capacitance is

(a) 0.25 μF

(b) 0.5 μF

(c) 1 μF

(d) 2 μF

74 Refer to Fig. Q.31. The resistance between terminals A and B of the network shown is

(a) 3.86 Ω

(b) 10.8 Ω

(c) 12 Ω

(d) 26 Ω

75 Refer to Fig. Q.32. Block X is

(a) an erase head

(b) a playback amplifier

(c) a high frequency bias

(d) a sync generator

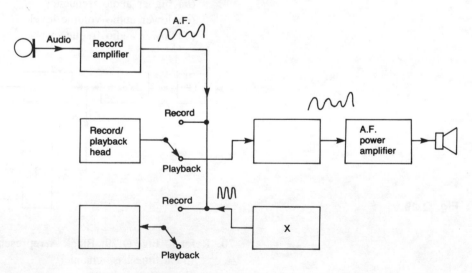

Fig. Q.32

76 In a high definition television system the number of scan lines is

(a) 50

(b) 425

(c) 625

(d) 1250

77 Figure Q.33 shows the input and output waveforms of a 2-input gate. The gate is
(a) AND
(b) OR
(c) Ex-OR
(d) NOR

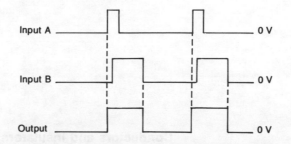

Fig, Q.33

78 Refer to Fig. Q.34. Which one of the following correctly identifies blocks A, B and C?

	A	B	C
(a)	display seconds	display minutes	display hours
(b)	display minutes	display seconds	display hours
(c)	display hours	display minutes	display seconds
(d)	display seconds	display hours	display minutes

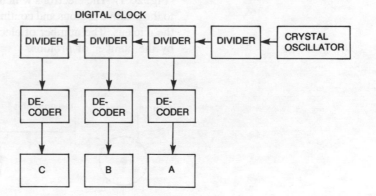

Fig. Q.34

79 Block X in Fig. Q.35 is
(a) an analogue-to-digital converter
(b) a digital-to-analogue converter
(c) a digital decoder
(d) a multiplexer

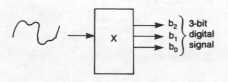

Fig. Q.35

80 NICAM provides
(a) hi-fi monophonic sound
(b) hi-fi stereophonic sound
(c) high quality TV picture
(d) improved time base synchronisation

20 The semiconductor diode

Conductors and insulators

All substances are made up of one or more chemical elements such as oxygen, sulphur, etc. The smallest part of an element is the atom. Atoms of different elements may combine to form a molecule of a substance, e.g. a molecule of water comprises two atoms of hydrogen and one atom of oxygen. In this way different substances are made up.

The atom itself is composed of smaller particles known as electrons revolving around a central nucleus containing one or more protons (Fig. 20.1). The electrons which are negatively charged are attracted to the positive protons and continuously rotate in orbits or shells round the nucleus. The number of electrons in an atom is exactly balanced by the number of protons.

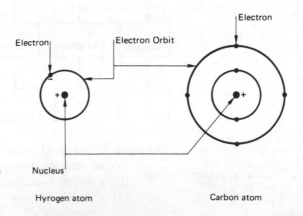

Fig. 20.1 Atoms consist of negative electrons rotating round a positive nucleus

Atoms of different elements are differentiated one from the other by the number of electrons it contains, e.g. a hydrogen atom has one electron while that of carbon has six. For electrical conduction to occur, electrons which are loosely connected to the nucleus (known as free electrons) leave their orbit when attracted by a positive potential, and begin to travel along forming a flow of electrons, or current.

A good conductor has a large number of 'loose' or free electrons. These free electrons will aid the flow of current. A good conductor will have such a small resistance that it can be neglected — examples are silver, copper and aluminium (Fig. 20.2).

An insulator is a material with very few free electrons. Insulators restrict the flow of current very heavily and thus have very large resistances approaching an open circuit. Examples are glass, dry wood, rubber, p.v.c., mica and polystyrene.

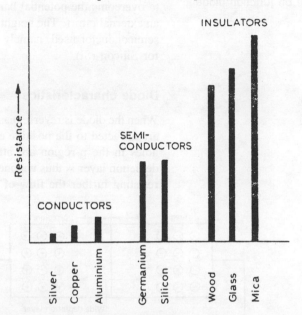

Fig. 20.2 Conductors, semiconductors and insulators

Semiconductors

Semiconductors have their atoms grouped together in a regular pattern called a 'crystal lattice'. They are not good conductors (hence their name) since they have few 'free electrons'. The number of 'free electrons' increases as the temperature goes up, which leads to improved conductivity. These 'free electrons' are known as minority carriers.

Conductivity may also be improved by adding controlled amounts of 'impurities'. Impurities such as arsenic atoms introduce extra electrons in the lattice structure producing the n-type semiconductor. Such atoms are known as *donor* atoms. On the other hand impurity atoms known as *acceptor* atoms, for instance aluminium atoms, introduce a 'shortage of electrons' referred to as holes and producing the p-type semiconductor (Fig. 20.3). The electrons and holes produced by the injection of impurities are known as the majority carriers.

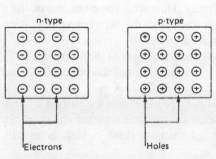

Fig. 20.3 n-type and p-type semi-conductors

n-region   p-region

Depletion layer

Fig. 20.4 The pn junction diode

The junction diode

When a p-type semiconductor is joined to an n-type semiconductor (Fig. 20.4), by a process known as *diffusion*, electrons from the n-region begin to cross over to fill the holes in the p-region. This movement continues until a neutral zone known as the *depletion layer* is established on either side of the pn junction. This depletion layer provides a potential barrier preventing further movement by electrons across the junction.

Electrons wishing to cross over must now have sufficient energy to overcome the potential barrier. Such energy may be provided by an external e.m.f. The height of the potential barrier depends on the semiconductor used, namely 0.3 V for Germanium (Ge) and 0.6 V for Silicon (Si).

Diode characteristics

When the diode is reverse biased (Fig. 20.5), electrons in the n-region are attracted to the positive electrode of the bias voltage, while the holes in the p-region are attracted to the negative electrode. The depletion layer is thus widened, producing a higher potential barrier resisting further the flow of electrons.

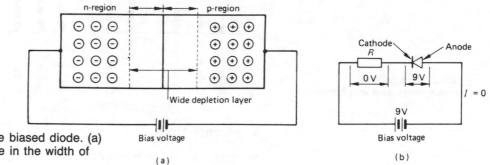

Fig. 20.5 Reverse biased diode. (a) Shows the increase in the width of the depletion layer

A forward bias, however, removes the depletion layer as shown in Fig. 20.6, allowing the electrons to cross over, with current due to majority carriers flowing freely. However, the diode maintains a constant voltage drop across it known as the forward voltage drop (0.3 V for Ge and 0.6 V for Si diodes).

The junction diode forward characteristics are as shown in Fig. 20.7. Note that as soon as the bias voltage overcomes the potential barrier of the diode, a large current begins to flow. A *very small* increase in the bias voltage produces a *very large* increase in the current through the diode. For bias voltages below the required forward voltage drop, a small leakage current (in microamps) flows which is usually neglected.

The reverse bias characteristics are as shown in Fig. 20.8. An extremely small current passes through the diode when in reverse bias,

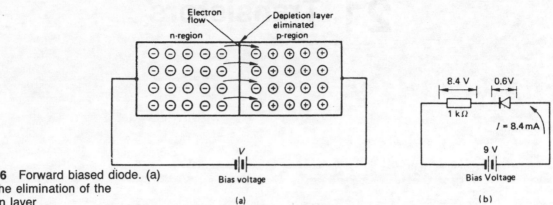

Fig. 20.6 Forward biased diode. (a) shows the elimination of the depletion layer

(a)

(b)

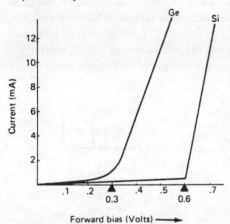

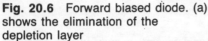

due to minority carriers. This reverse current is almost constant up to a maximum voltage known as the reverse breakdown voltage. If a higher voltage is applied, breakdown occurs and the reverse current increases rapidly causing damage to the diode. Care must always be taken to ensure that diodes are not subjected to reverse voltages higher than the breakdown voltage specified by the manufacturer. Germanium diodes have higher leakage currents and therefore lower reverse resistance than silicon diodes.

Fig. 20.7 Forward characteristics of germanium and silicon diodes

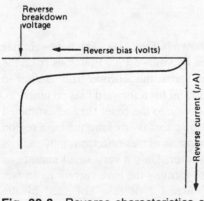

Fig. 20.8 Reverse characteristics of a junction diode

21 Transistors

The transistor is a semiconductor device made up of two pn junctions as shown in Fig. 21.1. There are three terminals to the transistor, known as the emitter, base, and collector. Transistors may be of two types: the pnp shown in (a) and the npn shown in (b). The principle of operation for both types is the same, the difference being in the d.c. bias conditions.

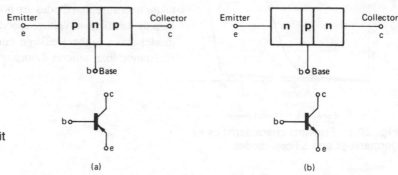

Fig. 21.1 (a) pnp transistor with it symbol (b) npn transistor with its symbol

Consider the npn transistor shown in Fig. 21.2. The base–emitter (b–e) junction is forward biased by voltage V_{BE}. Electrons from the emitter (I_e) will, therefore, flow across the junction from emitter to base. It is the normal forward current for a forward bias pn junction. As soon as the electrons cross over into the base, they are attracted by the positive potential of the collector. By making the base region very thin it is possible to have almost all of these electrons move across the base region towards the collector. Only a very small number of electrons is collected by the base forming the base current I_b. In fact 95% or more of the emitter current I_e is now collected by the collector to form the collector current I_c. Therefore,

$$I_e = I_c + I_b$$

Since base current I_b is very small (in μA) it is usually neglected. I_c and I_e are assumed to be equal and are generally referred to as the transistor current.

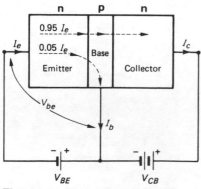

Fig. 21.2 Bias voltages for npn transistor

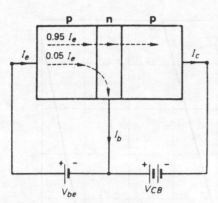

Fig. 21.3 Bias voltages for pnp transistor

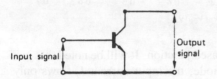

Fig. 21.4 Common emitter (CE) configuration

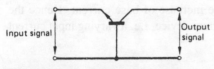

Fig. 21.5 Common base (CB) configuration

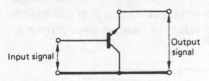

Fig. 21.6 Common collector (CC) configuration (Note that the output is taken from the emitter)

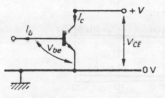

Fig. 21.7

It will be noticed that the base−collector (b−c) junction is reversed biased by voltage V_{CB}. This is essential otherwise the electrons will not be attracted towards the collector. Conventional flow of current from positive to negative will show transistor current flowing from collector to emitter.

For a pnp transistor the polarities of the d.c. supplies must be reversed as shown in Fig. 21.3. In this case the transistor current is a movement of holes from emitter to collector or electrons from collector to emitter.

Transistor configurations

There are three possible ways to connect a transistor when used in a circuit.

1. The **common emitter** (CE) configuration where the input signal is fed between the base and emitter, while the output is taken between collector and emitter (Fig. 21.4). This is the most widely used configuration due to its flexibility and high gain.
2. The **common base** (CB) configuration where the base is common to both input and output (Fig. 21.5).
3. The **common collector** (CC) configuration where the collector is common. This configuration is also known as the emitter follower (Fig. 21.6).

While the internal behaviour is exactly the same for each configuration, external behaviour is different for each case. Each configuration presents different characteristics in terms of its gain, input and output impedances and frequency response.

Common emitter characteristic curves

The transistor has three types of characteristic curve outlining its behaviour under static (or quiescent) conditions, i.e. with no input signal. These are:

1. The **input characteristics**, the graph of *input* current against *input* voltage.
2. The **output characteristics**, the graph of *output* current against *output* voltage.
3. The **transfer characteristics**, the graph of *output* current against *input* current.

The characteristics described below are for an npn transistor (Fig. 21.7). For a pnp transistor the d.c. voltages should be negative.

Input characteristics

Figure 21.8 shows the input characteristics for an npn transistor. It is the same as the characteristics of a forward biased pn junction since

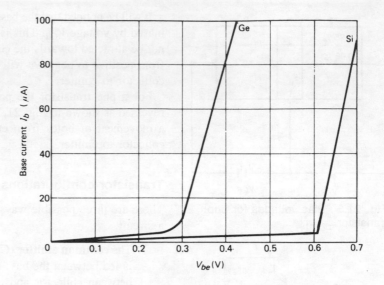

Fig. 21.8 Input characteristics of a transistor

the input (b−e) is a forward biased junction. It will be noted that just as in the case of a junction diode, the input current I_b flows only when the required forward voltage drop is established across the b−e junction. Once that voltage (0.3 V for Ge and 0.6 V for Si) is established, the voltage between base and emitter V_{be} remains practically constant for a large increase of base current. Hence the transistor is considered as a current device, i.e. a varying input current with a constant input voltage.

Output characteristics

Figure 21.9 shows a family of curves known as the output characteristics relating collector or output current I_c to collector or output voltage V_{CE} for specific values of base or input current I_b.

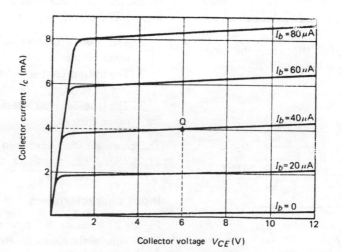

Fig. 21.9 Olutput characteristics of a transistor

These curves also provide a relationship between the input current on the one hand and output current and voltage on the other. For example, given a transistor with a base current of 40 μA and a collector voltage of 6 V, then using the output characteristics in Fig. 21.9 a collector current of 4 mA will be flowing.

The curve $I_b = 0$ represents the state of the transistor before it begins to conduct, i.e. at cut-off when V_{be} is below the required forward voltage drop. Theoretically, transistor current is zero when $I_b = 0$; however, a small leakage current does flow through to the collector.

Static current gain h_{fe}

A very important parameter of any transistor is its d.c. current gain, known as the static current gain h_{fe}. It is the current gain of the transistor at quiescence, i.e. without an input signal. It has no units (it is a ratio).

$$\text{Static current gain } h_{fe} = \frac{\text{Output current}}{\text{Input current}} = \frac{I_c}{I_b}$$

h_{fe} can be calculated from the output characteristics. For example, if the transistor is operating at point Q where $I_b = 40$ μA, and $I_c = 4$ mA then

$$h_{fe} = \frac{I_c}{I_b} = \frac{4 \times 1000}{40} = 100$$

Transfer characteristics

These are the relationships between the input and output currents (Fig. 21.10). The static current gain may also be calculated using this characteristic curve. For example, if Q is the operating point of the transistor then

$$h_{fe} = \frac{I_c}{I_b} = \frac{4 \times 1000}{40} = 100$$

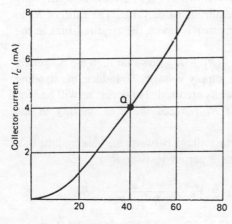

Fig. 21.10   Transfer (or mutual) characteristics of a transistor

22 The common emitter amplifier

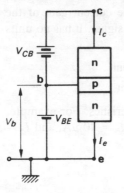

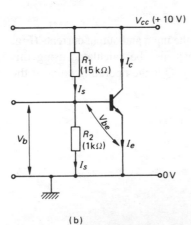

(a)

(b)

Fig. 22.1 Base bias for npn transistor

Before the transistor can be used as an amplifier it must have the correct d.c. bias, as shown in Fig. 22.1(a) for an npn transistor. It can be seen that the two voltages V_{BE} (providing the forward bias for the b−e junction) and V_{CB} (providing the reverse bias for the b−c junction) are connected in series. V_{BE} and V_{CB} can thus be substituted by the potential divider R_1−R_2 shown in Fig. 22.1(b). This way only one d.c. supply voltage is required, namely V_{cc}. The ratio of R_1 to R_2 is chosen so that the transistor is given the required bias at the base.

A standing current $I_s = V_{cc}/(R_1 + R_2)$ flows through the bias chain R_1−R_2 causing a drain on the supply voltage. To reduce this standing current, high values of R_1 and R_2 are used. However, as will be seen later, a very high value of R_1 reduces the d.c. stability of the transistor.

The voltage at the base is the voltage between the base and the 0 V line or chassis, i.e. the voltage across resistor R_2.

$$\text{Voltage at the base } V_b = V_{R_2} = \frac{V_{cc}}{R_1 + R_2} \times R_2.$$

For example, given $V_{cc} = 10$ V, $R_1 = 15$ kΩ, $R_2 = 1$ kΩ, then

$$V_b = \frac{V_{cc}}{R_1 + R_2} \times R_2 = \frac{10 \text{ V}}{15 \text{ k}\Omega + 1 \text{ k}\Omega} \times 1 \text{ k}\Omega$$

$$= \frac{10}{16} \times 1 = 0.625 \text{ V}$$

By using different values of R_1, R_2 or both, the base voltage may be varied.

The same biasing arrangement may be used for a pnp transistor as shown in Fig. 22.2. In this case the d.c. supply is negative, $-V_{cc}$. The function of bias chain $R_1 - R_2$ is the same as that for the npn transistor. The fact that the d.c. voltages are negative should be remembered but need not be included in calculation. Hence

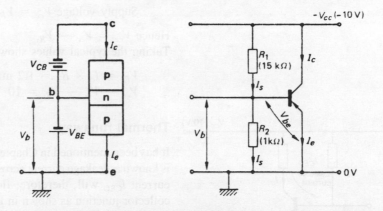

Fig. 22.2 Base bias for pnp transistor

$$V_b = \frac{V_{cc}}{R_1 + R_2} \times R_2 = \frac{10}{15 + 1} \times 1 = \frac{10}{16} = 0.625 \text{ V}$$

The voltage at the base is -0.625 V.

For a forward bias to be established across the b–e junction, the base voltage has to be 'higher' than the emitter voltage; 'higher', whether in the positive (npn) or in the negative (pnp) directions. In general, regardless of the type of transistor used, the base voltage is always higher than the emitter but lower than the collector.

As explained in the previous section, the current through the transistor is determined by the b–e forward bias V_{be}, i.e. the potential difference between base and emitter $V_{be} = V_b - V_e$. A variation in emitter or base voltage causes a change in transistor current. In the transistor circuit under consideration the emitter being at chassis potential, only the base can change with

$$V_{be} = 0.62 - 0 = 0.62 \text{ V}$$

For example, if the base voltage V_b goes away from the emitter (more positive for npn or more negative for pnp transistors), V_{be} increases causing the transistor current to go up. On the other hand if V_b goes towards the emitter, V_{be} decreases and the transistor current is reduced.

Collector or load resistor

In order to produce an output voltage at the collector, a load resistor R_3 (also known as the collector resistor) is added as shown in Fig. 22.3. The collector current I_c flows through collector resistor R_3 developing a voltage across it. Hence

$$V_{R_3} = I_c \times R_3$$

Since all voltages are measured with respect to chassis or earth potential, the collector voltage V_{ce} is the p.d. between the collector and the chassis. As can be seen from the circuit,

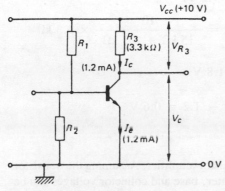

Fig. 22.3 The load resistor R_3

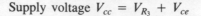

Supply voltage $V_{cc} = V_{R_3} + V_{ce}$

Hence $V_{ce} = V_{cc} - V_{R_3}$
Taking the typical values shown where $I_c = 1.2$ mA,

$$V_{R_3} = I_c \times R_3 = 1.2 \text{ mA} \times 3.3 \text{ k}\Omega = 4 \text{ V (approx.)}$$
$$V_{ce} = V_{cc} - V_{R_3} = 10 - 4 = 6 \text{ V}$$

Thermal runaway

It has been mentioned in Chapter 19 that minority carriers form what is known as leakage current across a reverse biased junction. A leakage current I_{CEO} will, therefore, flow across the reverse biased base–collector junction as shown in Fig. 22.4. It is amplified in the same way as the input (or base) current by a factor of h_{fe}. As the temperature of the transistor increases, so the leakage current goes up. This is amplified by the transistor, increasing the collector current which in turn raises the temperature further, causing the leakage current to go up, and so on. The process known as thermal runaway is cumulative and if unchecked could cause damage to the transistor.

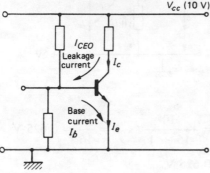

Fig. 22.4 Leakage current I_{CEO}

D.C. stabilisation

The effect of leakage current in the CE amplifier is to cause instability in d.c. or static operating conditions. This can be overcome by the addition of emitter resistor R_4 as shown in Fig. 22.5. The emitter voltage is now equal to the volt drop produced by the emitter current I_e flowing through emitter resistor R_4. Hence $V_e = I_e \times R_4$. D.C. stabilisation takes place as follows.

If I_c and I_e increase due to an increase in leakage current, the emitter voltage V_e increases with it. Since $V_{be} = V_b - V_e$ then an increase in V_e produces a decrease in V_{be}. This reduces the base current and restores I_c and I_e to their previous values. The emitter resistor R_4 provides a negative feedback loop giving stability to the static conditions of the amplifier. Using the typical values given in Fig. 22.5 and assuming emitter current $I_e = 1.2$ mA, then $V_e = I_e \times R_4 = 1.2 \text{ mA} \times 1 \text{ k}\Omega = 1.2 \text{ V}$.

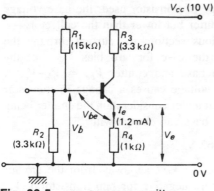

Fig. 22.5 npn common emitter amplifier with emitter resistor R_4

$$V_b = \frac{V_{cc}}{R_1 + R_2} \times R_2 = \frac{10}{15 \text{ k}\Omega + 3.3 \text{ k}\Omega} \times 3.3 \text{ k}\Omega$$

$$= \frac{10}{18.3} \times 3.3 = 1.8 \text{ V}$$

$$V_{be} = V_b - V_e = 1.8 - 1.2 = 0.6 \text{ V}$$

Using pnp transistors

Figure 22.6 shows a pnp transistor amplifier. Assuming the transistor to be silicon, its current, emitter, base and collector voltages can be calculated as follows:

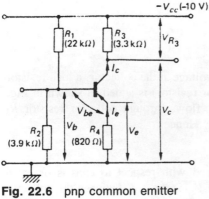
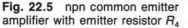

Fig. 22.6 pnp common emitter amplifier

$$V_b = \frac{V_{cc}}{R_1 + R_2} \times R_2 = \frac{10}{22\ \text{k}\Omega + 3.9\ \text{k}\Omega} \times 3.9\ \text{k}\Omega$$

$$= \frac{10}{25.9} \times 3.9 = 1.5\ \text{V}$$

From $V_{be} = V_b - V_e$ we get $V_e = V_b - V_{be}$ but $V_{be} = 0.6$ V (silicon transistor) and $V_b = 1.5$ V. Therefore

$$V_e = 1.5 - 0.6 = 0.9\ \text{V}$$

From $V_e = I_e \times R_4$, we get

$$I_e = \frac{V_e}{R_4} = \frac{0.9\ \text{V}}{820\ \Omega} = \frac{0.9\ \text{V}}{0.82\ \text{k}\Omega} = 1.1\ \text{mA}$$

$$I_c = I_e = 1.1\ \text{mA}$$
$$V_{R_3} = I_c \times R_3 = 1.1\ \text{mA} \times 3.3\ \text{k}\Omega = 3.6\ \text{V}$$
$$V_c = V_{cc} - V_{R_3} = 10 - 3.6 = 6.4\ \text{V}$$

The static operating conditions of the transistor are therefore

$$e - 0.9\ \text{V}, \quad b - 1.5\ \text{V}, \quad c - 6.4\ \text{V} \quad \text{with} \quad I_e = 1.1\ \text{mA}$$

The emitter, base and collector voltages given above are typical for a single stage amplifier such as an i.f. amplifier or a driver. The emitter is approximately $0.1 \times V_{cc}$, with the collector at approximately $0.6 \times V_{cc}$. It can be seen that for both types of transistor the emitter voltage is lowest while that of the collector is highest, with the base at approximately 0.6 V (for Si transistors) above the emitter.

The npn transistor with negative supply

It is possible to connect an npn transistor using a negative d.c. supply voltage $-V_{cc}$ as shown in Fig. 22.7. In this case the chassis is the positive rail. All voltages are, therefore, negative as they are all measured with respect to the positive chassis. Using the typical values given in the diagram we get

$$V_b = \text{voltage across } R_1 = \frac{V_{cc}}{R_1 + R_2} \times R_1$$

$$= \frac{10}{15 + 3.3} \times 15 = \frac{10}{18.3} \times 15 = 8.2\ \text{V}$$

$$V_c = \text{voltage across } R_3 = I_c \times R_3 = 1.2\ \text{mA} \times 3.3\ \text{k}\Omega$$
$$- 4\ \text{V (approx.)}$$
$$V_e = V_{cc} - V_{R_4} = V_{cc} - I_e \times R_4 = 10 - 1.2 \times 1$$
$$= 10 - 1.2 = 8.8\ \text{V}.$$

The static operating condition of the transistor is therefore: $e - 8.8$ V, $b - 8.2$ V, $c - 4$ V.

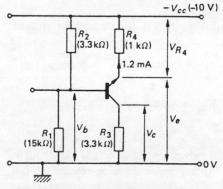

Fig. 22.7 npn transistor amplifier using negative supply $-V_{cc}$

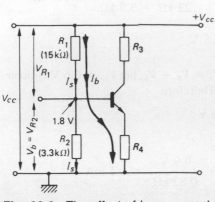

Fig. 22.8 The effect of base current I_b

Base current

Base current I_b (Fig. 22.8) flows from the positive d.c. rail through R_1 across the b–e junction to the emitter. Bias resistor R_1 thus has two currents flowing through it: standing current I_s (which also flows through R_2) plus the base current I_b (which does not flow through R_2). The volt drop across R_1 thus increases by $I_b \times R_1$. Since $V_{R_1} + V_{R_2} = V_{cc}$ then an increase in V_{R_1} would result in an equal decrease in V_{R_2}, i.e. a drop in base voltage. Under normal circumstances I_b is very small and may be neglected with V_b assumed to be determined by resistor chain $R_1 - R_2$.

However, if the base current is large (e.g. when a transistor is conducting very heavily) or if R_1 is made very large, the change in base voltage due to I_b begins to affect the static conditions of the transistor and has to be taken into account.

Consider the circuit in Fig. 22.8. Under normal conditions a base current of say 10 μA produces a voltage drop across R_1 of $I_b \times R_1$ $= 10 \times 10^{-6} \times 15 \times 10^3 = 150 \times 10^{-3} = 0.15$ V which is small compared with a base voltage of 1.8 V determined by bias chain $R_1 - R_2$. If the transistor is now made to conduct very heavily, taking a large current, base current will also be high, say 80 μA. Voltage across R_1 due to 80 μA base current is $80 \times 10^{-6} \times 15 \times 10^3 = 80 \times 15 \times 10^{-3} = 1.2$ V. The base voltage will drop by an equal amount, i.e. from 1.8 V to 0.6 V.

Base current biasing

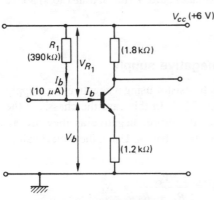

Fig. 22.9 Base current biasing

The base current can be used to provide the normal bias for a transistor as shown in Fig. 22.9. In this circuit R_2 is dispensed with and a very large R_1 is used. Current I_b is now entirely responsible for the volt drop across R_1 (there is no standing current). This volt drop is large enough to provide the normal bias.

Given that the base current in Fig. 22.9 is 10 μA the base voltage may be calculated as follows:

$$V_{R_1} = I_b \times R_1 = 10 \times 10^{-6} \times 390 \times 10^3 = 3900 \times 10^{-3}$$
$$= 3.9 \text{ V}$$

The voltage at the base is the voltage between the base and chassis, i.e.

$$V_b = V_{cc} - V_{R_1} = 6 - 3.9 = 2.1 \text{ V}$$

The circuit in Fig. 22.9 has the advantage of having a higher input impedance due to the absence of the input shunting resistor R_2 but it has no d.c. stability.

Cut-off and saturation

A transistor is said to be at cut-off when it ceases to conduct, i.e. its current is zero. With $I_e = 0$, no voltage is developed across R_4

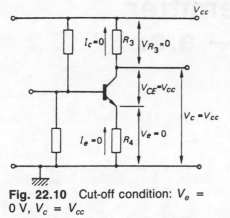

Fig. 22.10 Cut-off condition: $V_e = 0$ V, $V_c = V_{cc}$

(Fig. 22.10). Hence the emitter voltage ($V_e = I_e \times R_4$) is also zero. A zero collector current produces no p.d. across R_3 ($V_{R_3} = I_c \times R_3$) making the collector go to the supply V_{cc}. Voltage between collector and emitter V_{CE} (= $V_c - V_e$) is, therefore, equal to V_{cc}.

A transistor is said to be saturated when the current through it is so high that it cannot increase any further, i.e. when I_e and I_c are at their maximum value. As I_e increases, V_e also increases (Fig. 22.11). The voltage across R_3 increases as I_c increases, forcing the collector to go away from V_{cc} towards the emitter ($V_c = V_{cc} - V_{R_3}$). Hence as the current through the transistor increases, the emitter voltage goes towards the collector while the collector goes towards the emitter. At saturation, when the transistor current is at a maximum, V_e and V_c are approximately equal, with V_{CE} almost at zero. Figure 22.11 gives typical values for saturation.

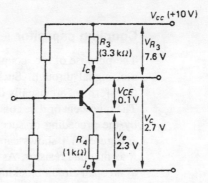

Fig. 22.11 Saturation condition: $V_e \cong V_c$

A transistor may therefore be used as a **switch** (Fig. 22.12):

switch CLOSED — transistor saturated
switch OPEN   — transistor cut-off

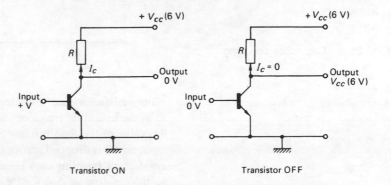

Fig. 22.12 The transistor as a switch

23 Common emitter amplifier — a.c. operation

Coupling capacitor

The purpose of a.c. coupling is to block the d.c. while allowing the a.c. signal through. Such coupling devices, e.g. a capacitor or a transformer, are usually used at the input and output of an amplifier. The quiescent or d.c. conditions of the transistor are thus not affected by the preceding or succeeding stages.

Figure 23.1 shows capacitor C coupling point A to point B with R as the load resistor. As far as d.c. is concerned, the capacitor acts as an open circuit blocking the d.c. between A and B. For this reason a coupling capacitor is also known as a *blocking capacitor*.

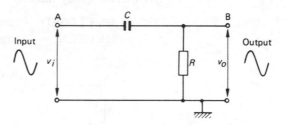

Fig. 23.1 Capacitor coupling

For satisfactory a.c. coupling, the capacitor must have a reactance X_c at the operating frequency which is very small as compared with the value of the load resistor R so that a very small fraction of the input signal is dropped across C. For instance if $V_1 = 100$ mV, then satisfactory coupling may be said to be achieved if the output $V_o = 95$ mV, a drop of 5 mV (5%) due to the coupling capacitor.

The value of a coupling capacitor depends on two factors.

1. *Load resistor R*. Assuming that satisfactory coupling is achieved when $X_c = R/20$ and given that $R = 1$ kΩ and the operating frequency $f = 300$ Hz then,

$$X_c = \frac{1}{20} R = \frac{1}{20} \times 1000 = 50 \ \Omega$$

From $X_c = 1/2\pi fC_1$, we get

$$C_1 = \frac{1}{2\pi fX_c} = \frac{1}{2\pi \times 300 \times 50}$$

$$= 10.6 \ \mu\text{F o4 } 10 \ \mu\text{F approx.}$$

If the load resistor is increased to 100 kΩ,

$$X_c = \frac{1}{20} R = \frac{1}{20} \times 100 \text{ k}\Omega = 5 \text{ k}\Omega$$

$$C_2 = \frac{1}{2\pi fX_c} = \frac{1}{2\pi \times 300 \times 5 \times 10^3} = 0.1 \ \mu\text{F}$$

Hence if the load resistor is increased by 100 times (from 1 kΩ to 100 kΩ), the coupling capacitor may be decreased by the same proportion (from 10 μF to 0.1 μF).

In general the higher the load resistor, the smaller the required value of the coupling capacitor.

2. *The frequency of operation.* Take the above example where satisfactory coupling was achieved when $C = 10 \ \mu$F and $R = 1$ kΩ with $f = 300$ Hz.

If the frequency is now increased to 300 kHz then with $X_c = R/20$ = 50 Ω

$$C_3 = \frac{1}{2\pi fX_c} = \frac{1}{2\pi \times 300 \times 10^3 \times 50} = 0.01 \ \mu\text{F}$$

Hence if the frequency is increased by 1000 times (from 300 Hz to 300 kHz), the coupling capacitor may be decreased by the same proportion (from 10 μF to 0.01 μF).

In general for a given value of R a large coupling capacitor is required for low frequencies and vice versa.

For a range of frequencies the value of the coupling capacitor is determined by the lowest frequency. Considering the above example, a 10 μF capacitor calculated to provide adequate coupling at 300 Hz will provide more than adequate coupling at 300 kHz. On the other hand a 0.1 μF capacitor calculated to provide adequate coupling at 300 kHz does not provide anything like adequate coupling at 300 Hz.

Decoupling

Figure 23.2(b) shows a capacitor C decoupling a resistor R. Without the capacitor (Fig. 23.2(a)) point A has a d.c. voltage of 10 V and an a.c. signal voltage of 10 mV. The capacitor being open circuit for d.c. does not therefore interfere with the d.c. voltage at point A. However, if the value of C is such that at the operating frequency

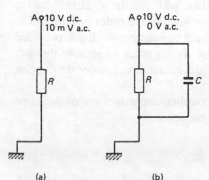

(a)   (b)

Fig. 23.2 Effect of decoupling capacitor (a) Without decoupling capacitor (b) With decoupling capacitor

its reactance is very small compared with R, all the a.c. signal will effectively be short-circuited to earth. Point A will thus have zero signal voltage. The value of C that would provide satisfactory decoupling is determined by resistor R as well as the operating frequency, in the same way as the value of the coupling capacitor described above.

R–C coupled amplifier

Figure 23.3 shows an $R-C$ coupled amplifier where C_1 is the input coupling capacitor. The value of C_1 is comparatively large due to the low input impedance of the CE transistor which is made even smaller by R_2 shunting the input. C_2 couples the output to the load or the next stage and is of a similar value to C_1. Typical values of coupling capacitors are as follows:

for a.f., 10 μF – 50 μF; for r.f., 0.01 μF – 0.1 μF.

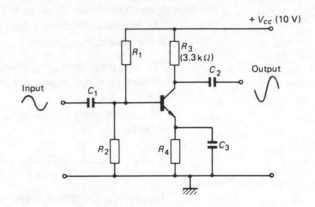

Fig. 23.3 $R-C$ coupled amplifier with decoupling capacitor C_3

Decoupling capacitor

The negative feedback produced by R_4 in Fig. 23.3 on the one hand provides the necessary d.c. stability, while on the other reduces the gain of the amplifier to a very low level of the order of 2 to 3. The reduction in the gain is due to the a.c. negative feedback provided by the signal voltage drop across R_4. In order to remove the a.c. negative feedback while retaining d.c. stability, emitter decoupling capacitor C_3 is used.

Typical values for emitter decoupling capacitors are of the same order as those for the coupling capacitor.

Amplification

The circuit in Fig. 23.3 is the complete circuit for a single stage CE amplifier. When an input signal, e.g. a sine wave, is fed as shown C_1 couples the signal to pass through to the base. For the positive half cycle, the base goes away from the emitter, V_{be} goes up, I_e and

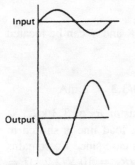

Fig. 23.4 Phase reversal

hence I_c increase, bringing V_c down. Thus the positive half cycle of the input produces a negative-going half cycle at the output. On the other hand the negative half cycle of the input produces an upward or a positive movement in the collector voltage. The output is thus out of phase with the input as shown in Fig. 23.4. Amplification occurs because a very small swing in V_{be} produces a large swing in transistor current which, when passing through R_3, produces a large voltage swing at the collector.

The load line

Transistor output characteristics show the general operation of the transistor. In order to represent the operation of the transistor when used in a circuit, a load line is drawn. Figure 23.5 shows the output characteristics for the transistor used in Fig. 23.3 with XY as the load line.

Before the load line can be drawn, two points falling on it have to be located. The most convenient points are point X on the x-axis where $I_c = 0$ and point Y on the y-axis where $V_c = 0$. These two points are then joined together to form the load line. Assume $V_C = V_{CE}$.

Point X. At this point, transistor current I_c is zero. The transistor is at cut-off. Hence the collector voltage $V_c = V_{cc}$.

Point Y. Here collector voltage $V_c = 0$. Putting $V_c = 0$ in the equation $V_{cc} = V_c + V_{R_3}$, we get $V_{cc} = V_{R_3}$. But $V_{R_3} = I_c \times R_3$, therefore $V_{cc} = I_c \times R_3$, hence

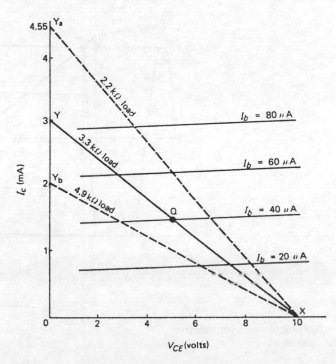

Fig. 23.5 The load line

$$I_c = V_{cc}/R_3$$

Using the values given in Fig. 23.3, points X and Y can be located as follows:

Point X. $I_c = 0$, $V_c = V_{cc} = 10$ V.
Point Y. $V_c = 0$, $I_c = V_{cc}/R_3 = 10/3.3 = 3$ mA

XY is therefore the load line for a load resistor R_3 of 3.3 kΩ.

Using a smaller load resistor, 2.2 kΩ, the load line is shifted to XY$_a$. Point X is the same as for the previous case since V_{cc} remains at 10 V. Point Y$_a$ however, is at $I_c = V_{cc}/R_3 = 10$ V/2.2 kΩ = 4.55 mA.

On the other hand a higher load resistor, e.g. 4.9 kΩ, would produce a load line XY$_b$ with point Y$_b$ at $I_c = 10$ V/4.9 kΩ = 2 mA.

Graphical analysis

Amplification of a signal takes place along the load line and may be represented as shown in Fig. 23.6. Point Q is the quiescent point representing the no-signal or d.c. operating point of the amplifier. At the Q-point the transistor is biased to have a base current

$$I_b = 20 \ \mu\text{A with } I_c = 1.5 \text{ mA and } V_c = 5 \text{ V}.$$

When a signal is applied, the base current swings as shown between

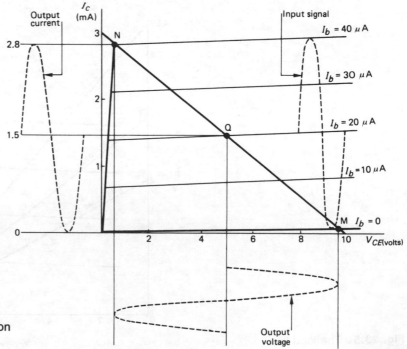

Fig. 23.6 Graphical representation
of amplifier operation

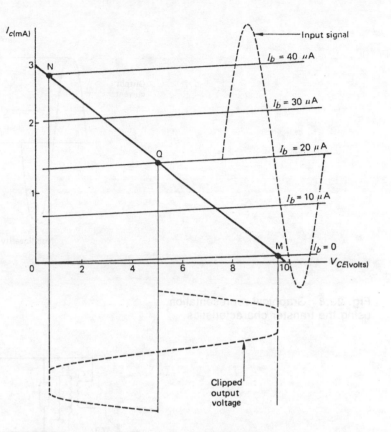

Fig. 23.7 Overdriven amplifier producing clipped output

0 μA and 40 μA. This produces a collector output current I_c swing of 2.8 mA and a voltage swing of approximately 9 V.

The input swing is restricted on the one hand by the line $I_b = 0$ when the transistor is at cut-off (point M on the load line) and on the other by line $I_b = 40$ μA when the transistor is saturated (point N on the load line). In the case of the amplifier under consideration the Q-point is chosen in the middle so that by feeding a signal of a peak value of 20 μA into the base, the base current swings between zero and 40 μA producing the maximum undistorted output swing shown. Any attempt to enlarge the input signal will result in a distorted output waveform as shown in Fig. 23.7, which is the case of an overdriven amplifier producing clipped sine waves. The input and output waveforms may be represented graphically using the transfer characteristics of Fig. 23.8. The operation is restricted to the linear part, otherwise distortion occurs.

Transformer coupled amplifier

Interstage coupling can be achieved by means of a transformer as shown in Fig. 23.9. R_1 and R_2 form a bias chain for TR_2 (the bias chain for TR_1 is not shown). C_1 is the bias decoupling capacitor which prevents any signal from developing across bias resistor R_2.

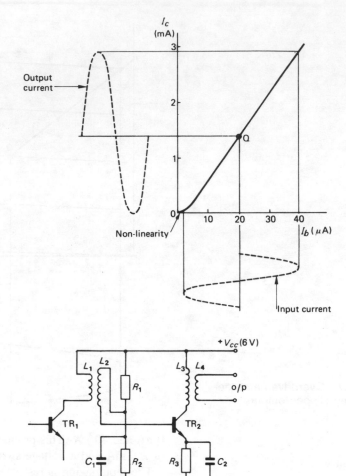

Fig. 23.8 Graphical representation using the transfer characteristics

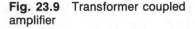

Fig. 23.9 Transformer coupled amplifier

R_3 is the emitter resistor and C_2 is the emitter decoupling capacitor. The primary windings of the two transformers L_1 and L_3 act as the load for TR_1 and TR_2 respectively. Since transformer windings have a very small resistance, the d.c. voltage at the collector is at a constant value of V_{cc}. The signal develops across the winding due to its a.c. resistance, i.e. its reactance. Two load lines can thus be drawn as shown in Fig. 23.10. If the supply voltage is 6 V, then the d.c. line is a vertical line representing a constant voltage at the collector of 6 V. The a.c. line represents the operation of the amplifier as the signal is applied. The point of intersection is the quiescent point Q.

Comparison between R–C and transformer coupled amplifiers

$R–C$ coupling is used extensively in electronics due to its high gain, better frequency response, cheap and small-size components.

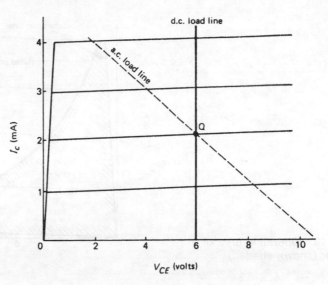

Fig. 23.10 a.c. and d.c. load lines for transformer coupled amplifier in Fig. 23.9

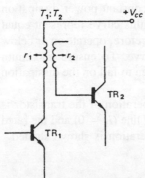

Fig. 23.11 Transformer matching = $r_1/r_2 = T_1^2/T_2^2 = n^2$

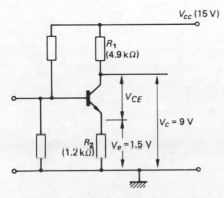

Fig. 23.12 Power dissipation in a transistor = $I_c \times V_{CE}$

Transformers, on the other hand, are large in size and more expensive. Transformer coupling has two main advantages:

1. It does not use a collector resistance, with the result that power dissipation, i.e. loss in the form of heat, is small compared with the $R-C$ coupled amplifier. Transformer coupled amplifiers are, therefore, more efficient.

2. It facilitates matching between stages (see Chapter 8). If r_1 is the output impedance of TR_1 and r_2 the input impedance of TR_2 (Fig. 23.11), then as was explained in Chapter 8

$$\frac{r_1}{r_2} = n^2 \text{ where } n \text{ is the turns ratio } \frac{T_1}{T_2}$$

Power dissipation

Transistors like other components have a specific power rating which must not be exceeded otherwise the transistor might be damaged.

The power dissipated in a transistor is $I_c \times V_{CE}$ watts.

For example, referring to Fig. 23.12 with $V_c = 9$ V, $V_e = 1.5$ V.

$$V_{CE} = V_c - V_e = 9 - 1.5 = 7.5 \text{ V}$$

$$I_c = I_e = \frac{V_e}{R_2} = \frac{1.5 \text{ V}}{1.2 \text{ k}\Omega} = 1.25 \text{ mA}$$

Therefore power dissipation in the transistor is

$$I_c \times V_{CE} = 1.25 \text{ mA} \times 7.5 \text{ V} = 9.4 \text{ mW}$$

Note that power dissipated in load resistor R_1 is

$$I_c^2 \times R_1 = (1.25 \text{ mA})^2 \times 4.9 \text{ k}\Omega = 7 \text{ mW}$$

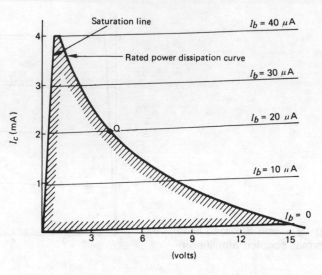

Fig. 23.13 Operating area of a transistor (shown shaded)

The power rating of a transistor may be represented on the output characteristic curve by a hyperbola or a constant power dissipation curve as shown in Fig. 23.13. Points on this curve represent equal power dissipation. The transistor must, therefore, operate on or below its maximum rated power dissipation curve. To ensure maximum power output, the Q-point is usually chosen to fall on the dissipation curve.

As can be seen from Fig. 23.13, the operation of the transistor is restricted by the saturation line, the cut-off line ($I_b = 0$) and the rated power dissipation curve. The area of operation is shown shaded.

24 The common base and common collector amplifiers

The CB amplifier

Figure 24.1 shows an amplifier using a transistor in the common base configuration. Two separate e.m.f.s are used to provide the necessary bias. C_1 is a coupling capacitor feeding the input between the emitter and the base. The output is taken across the load resistor R_4.

Figure 24.2 shows a practical i.f. amplifier using one d.c. supply. The load is a tuned circuit $C_3 - L_1$ employing transformer coupling, C_1 is the input coupling capacitor, $R_1 - R_2$ forms the bias chain, and R_3 is the emitter resistor. C_2 is a decoupling capacitor ensuring that the base is at zero a.c. potential. In this case, the decoupling capacitor is taken to the supply line instead of the chassis. This is permissible since the supply line is at zero potential as far as a.c., i.e. the signal, is concerned. The difference between the supply line and the chassis is a d.c. potential only.

The CB amplifier has a low input impedance (50−100 Ω) and a

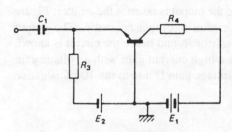

Fig. 24.1 CB amplifier with E_1 and E_2 providing the bias voltages

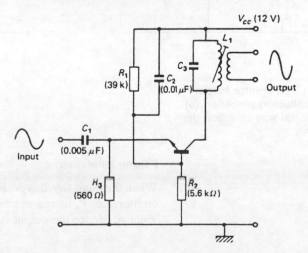

Fig. 24.2 Typical i.f. amplifier employing CB configuration

low gain compared with the CE configuration. It has the advantage, however, of a good high frequency response. Hence it is used at very high frequencies, e.g. r.f. amplifiers in radio and TV receivers, i.f. amplifiers for f.m. receivers, etc.

Phase relationship

If the input signal is negative-going, the emitter goes away from the base, and I_c increases to produce a drop in the collector of output voltage. The output is, therefore, in phase with the input.

The CC amplifier

Figure 24.3(a) shows a circuit using a transistor in the common collector configuration. C_1 and C_2 are the input and output coupling capacitors and $R_1 - R_2$ forms the bias chain for the base voltage. No collector resistor is used since the output is taken at the emitter. Figure 24.3(b) shows the way the circuit is drawn normally. The output develops across the emitter resistor R_3 and hence the circuit is known as an *emitter follower*. It has a high current gain with a voltage gain of just below one. The low voltage gain is due to the 100% negative feedback produced by R_3.

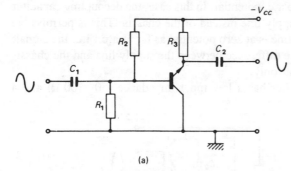

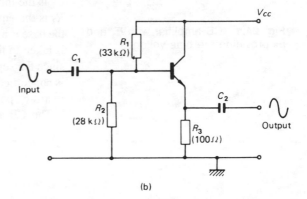

(a)   (b)

Fig. 24.3 (a) The emitter follower (or common collector) amplifier; (b) shows the normal way in which the circuit is drawn

Phase relationship

When the input signal is positive-going, the base goes away from the emitter, i.e. V_b increases producing an increase in emitter or output voltage. Hence the output is in phase with the input.

Comparison between CE, CB and CC configurations

Config-uration	Input impedance	Output impedance	Phase reversal	Advantages
CE	1–2 kΩ	10–50 kΩ	Yes	High voltage and power gain
CB	Very low	Very high	No	Good high frequency response
CC	Very high	Very low	No	(a) Low output impedance (b) High current gain

25 Photoelectric devices

Types of photoelectric devices

The **photo-emissive cell** or phototube, which employs the principle of electron emission from a cathode when light falls upon its surface.

The **photoconductive cell**, which reduces its resistance as the intensity of light falling upon it is increased.

The **photovoltaic cell** which generates an e.m.f. according to the intensity of light falling upon its surface. Semiconductor material such as selenium may be used; it produces a high enough voltage to deflect a voltmeter and can be used in lightmeters. The photovoltaic cells are also used to provide power for telephone repeaters and for communication satellites. They form the basis of solar generators where sunlight is used to generate electric power.

Photodiodes and **phototransistors**. In the photodiode a pn junction is reverse biased. The only current passing through the junction is, therefore, a small leakage current, which is a nuisance when the junction is used normally. In the photodiode it becomes the working current. It was found that the leakage current not only increases with heat but also with light waves falling upon the junction. That is why diodes and transistors are packed inside opaque containers. Photodiodes on the other hand are provided with a small window through which light can fall on the junction increasing its leakage current.

Phototransistors use the same principle as the photodiode. The junction in this case is the reverse biased base—collector junction. The advantage of the phototransistor is that the increased b—c leakage current due to light is amplified by the transistor which may then provide enough power to deflect a voltmeter as in the case of a lightmeter. Figure 25.1 shows a simple phototransistor circuit. Multiple phototransistors may be used to drive a motor.

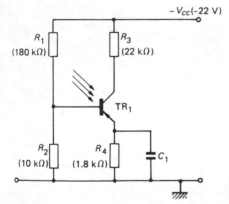

Fig. 25.1 Circuit using a photo-transistor

The light emitting diode

In a forward biased pn junction, electrons travelling from the n-region are injected into the p-region. Some of these electrons combine with the holes in the p-region. These combinations result in a release of

Fig. 25.2

energy in the form of visible light. For a large and efficient energy release, materials such as gallium or arsenic are used instead of silicon or germanium as the semiconductor material. Such semiconductor devices are known as light emitting diodes (LED) (Fig. 25.2). The wavelength of the light emitted, i.e. its colour, depends on the material used. Red and green emission are the most popular colours normally used.

Opto-couplers or isolators

An important use of light emitting diodes and photo-sensitive devices such as the photodiode or the phototransistor is in opto-coupling. In its simplest form, an opto-coupler or isolator consists of an LED emitting infra-red light which is picked up by a photo-sensitive device. Both the LED and the photo-sensitive device are enclosed in a single package (Fig. 25.3). Opto-coupling is superior to transformer coupling in switching and chopping applications. It has greater immunity to noise, hum, transients and bounce. Signal transfer is unidirectional with the result that changes in the load conditions have no effect on the input. Opto-coupling is also faster than isolation transformers or relays.

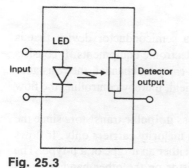

Fig. 25.3

26 Field effect transistors

The field effect transistor (fet) is a semiconductor device that is becoming very popular in modern electronic equipments. It consists basically of a semiconductor current-carrying channel, the resistance of which is controlled by an electric field, thereby controlling the flow of current through the channel.

Field effect transistors are known as **unipolar** transistors since the flow of current is carried out by the majority carriers only. It flows through one type of semiconductor, either an n-type or a p-type. The action of the normal junction transistor on the other hand is based on minority as well as majority carriers. This is because its current flows through a forward biased b−e junction (majority carriers) and a reverse biased b−c junction (minority carriers). Hence they are known as **bipolar** transistors.

Field effect transistors have three electrodes known as the source (s), the gate (g), and the drain (d). They correspond to the emitter, base and collector for the bipolar transistor.

Field effect transistors require no heating and are small in size. They have very high input impedance. They are less temperature-sensitive than the bipolar transistors and hence less prone to thermal runaway. Circuits using fets are simple in design, using fewer components than an equivalent circuit using bipolar transistors.

Field effect transistors are simple to manufacture, and more amenable for use in integrated circuits than their counterparts, the bipolar transistors.

There are two main types: the junction field effect transistor and the MOS field effect transistor.

The junction fet

Consider an n-type semiconductor channel with a d.c. potential V_{DD} connected across it as shown in Fig. 26.1(a). A current known as drain current I_D flows through the channel from drain to source as shown. If now a p-type region known as the gate is diffused into the n-type channel as shown in Fig. 26.1(b), a pn junction is formed. At the junction a depletion layer is established in the same manner

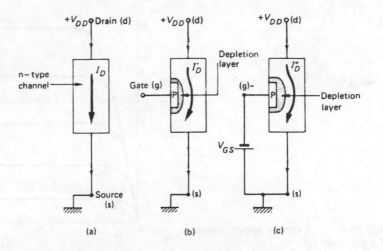

Fig. 26.1 The operation of a junction fet

(a)   (b)   (c)

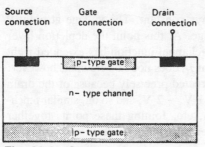

Fig. 26.2 Cross-section of the structure of a junction fet

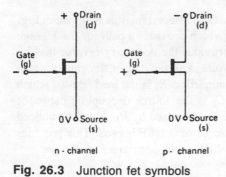

Fig. 26.3 Junction fet symbols

as a normal pn junction. As can be seen, the depletion layer restricts the flow of current through the channel by reducing the effective width of the channel. In other words it increases the resistance of the channel. The depletion layer may be widened, thus restricting further the flow of current by applying a reverse bias V_{GS} to the junction as shown in Fig. 26.1(c). By changing the bias voltage at the gate, drain current I_D may be varied. Figure 26.2 shows a cross-section of the structure of a field effect transistor.

FETs with p-type channel may also be used, with negative d.c. supply $-V_{DD}$. The symbols for both types of junction fets are shown in Fig. 26.3.

Output characteristics

The output characteristics for a junction fet used in the common source configuration is shown in Fig. 26.4. They are similar to the output characteristics of a bipolar transistor. They show how output current I_D changes with output voltage (voltage between drain and source) V_{DS} for specific values of gate bias (voltage between gate and source) V_{GS}.

The change in gate–source bias is large (in the region of a few volts) unlike the bipolar transistor where the base–emitter voltage is basically constant.

It can be seen that as the gate bias increases, so drain current decreases, until the depletion layer due to the gate–channel junction fills up the whole width of the channel, stopping the flow of current. The fet is then said to be cut off.

Pinch-off voltage

Consider the output curve with $V_{GS} = 0$ (Fig. 26.4). As the drain voltage V_{DS} is increased from zero, drain current increases gradually

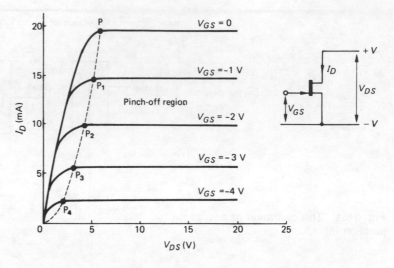

Fig. 26.4 The output characteristics of a junction fet

until point P, after which it stays constant. The voltage at point P is known as the pinch-off voltage. At this point the depletion layer due to the reverse biased gate—channel junction has extended right across the channel. Drain current I_D does not stop flowing, however, because the depletion layer is created precisely because of the drain current itself. For grid bias of -1 V, -2 V, and so on, similar pinch-off points are observed P_1, P_2, etc. Joining these points together forms the pinch-off region where the fet is made to operate.

The fet common source amplifier

A typical a.f. amplifier using a field effect transistor is shown in Fig. 26.5. R_1 is a leakage resistor which provides a path for the leakage current to go to chassis; R_3 provides the necessary reverse bias by giving the source a positive voltage above that of the gate. R_3 also provides d.c. stability for the amplifier. R_2 is the load resistor which may be as high as 1.5 MΩ. C_2 is the source decoupling capacitor removing the a.c. negative feedback caused by R_3. It will be noticed that a low value coupling capacitor of 0.1 μF is used. This provides adequate coupling due to the high input impedance of the fet.

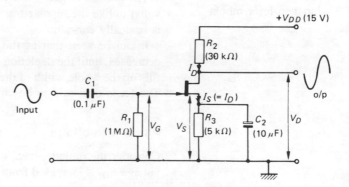

Fig. 26.5 a.f. amplifier using an n-channel junction fet

When a signal is applied into the input, the fet drain current will vary accordingly, producing an output voltage at the drain. For the positive half cycle, the gate goes in the positive direction, thereby reducing the reverse bias of the gate–source junction, and thus increasing the fet current I_D. As I_D increases, drain output voltage decreases and a negative-going half cycle is produced. On the other hand, the negative half cycle of the input produces a positive-going half cycle at the output. Hence the output is in anti-phase to the input.

Calculations

One advantage of the field effect transistor is its very low leakage current which is in the region of a few pico-amperes (10^{-12} mA). Hence the gate is at zero potential. Current through the fet flows from drain to source and it is usually referred to as drain current I_D (= source current I_S).

Consider the circuit in Fig. 26.5. Given a drain current I_D = 0.2 mA then source voltage V_S = 0.2 mA $\times$ 5 kΩ = 1 V, giving a reverse bias at the gate of 1 V.

Voltage across R_2 = 0.2 mA $\times$ 30 kΩ = 6 V.

Therefore drain voltage V_D = 15 − 6 = 9 V.

The load line

A load line may be drawn in the same way as that for a bipolar transistor. Figure 26.6 shows the load line for the amplifier in Fig. 26.5.

When I_D = 0, $V_{DS} = V_{DD}$ = 15 V giving us point X on the load line.

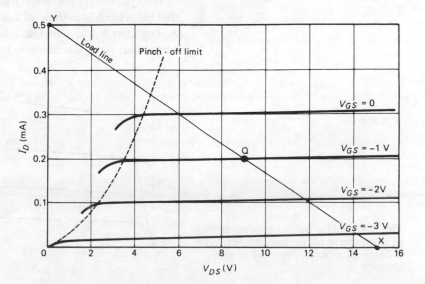

Fig. 26.6 Load line for the fet amplifier in Fig. 26.5

When $V_{DS} = 0$, the total d.c. supply V_{DD} falls across R_2.

Hence $I_D = \dfrac{V_{DD}}{R_2} = \dfrac{15\ \text{V}}{30\ \text{k}\Omega} = 0.5$ mA giving us point Y.

The quiescent point Q is chosen so that the transistor operates within its pinch-off region.

Quiescent or operating point Q is hence at $I_D = 0.2$ mA and $V_{GS} = -1$ V with $V_{DS} = 9$ V.

MOSFET

This type of field effect transistor has a metal gate which is electrically insulated from the semiconductor, by a thin oxide film. Hence its name MOS, which stands for Metal Oxide Silicon.

The n-type channel is formed by the gate insulating oxide attracting electrons from a p-type substrate (Fig. 26.7). The thickness of the n-channel may be varied by applying a voltage to the gate. A positive voltage widens the n-type channel, increasing the current, while a negative voltage narrows the n-channel and reduces the current. The opposite is true for a p-channel type.

There are two types of mosfets: the enhancement (normally off) and the depletion (normally on). In the **enhancement** type the fet is at cut-off (normally off), when gate bias $V_{GS} = 0$. Current flows only when a bias voltage is applied to the gate. The output characteristics for an n-channel enhancement type are shown in Fig. 26.8, together with its symbol.

The **depletion** type conducts (normally on) without a gate bias. The gate voltage may be positive or negative. For an n-channel type, the drain current increases as gate voltage goes positive and decreases as gate voltage goes negative (Fig. 26.9).

The symbols for p-channel mosfets are shown in Fig. 26.10. Note that the broken lines indicate a normally off (i.e. enhancement) type. A continuous line, on the other hand, indicates a normally on (depletion) type. Contact b is the substrate lead which is normally

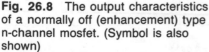

Fig. 26.7 Cross-section of a mosfet

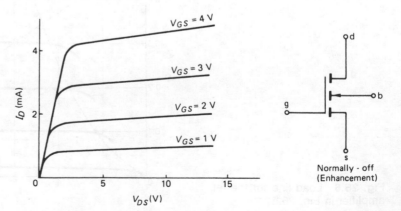

Fig. 26.8 The output characteristics of a normally off (enhancement) type n-channel mosfet. (Symbol is also shown)

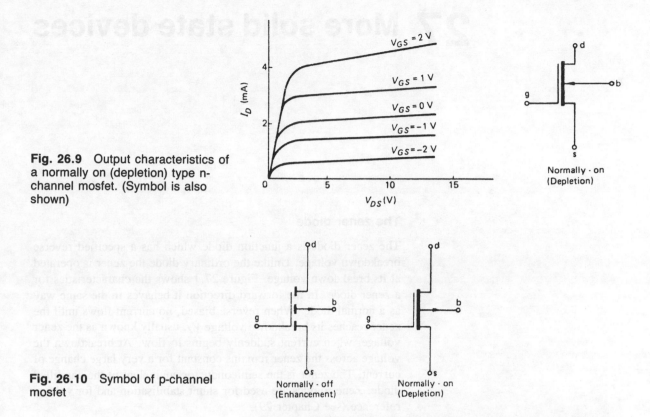

Fig. 26.9 Output characteristics of a normally on (depletion) type n-channel mosfet. (Symbol is also shown)

Normally - on
(Depletion)

Fig. 26.10 Symbol of p-channel mosfet

Normally - off
(Enhancement)

Normally - on
(Depletion)

taken to source potential. Figure 26.11 shows a typical amplifier using a p-channel depletion mosfet in the common source configuration. A negative d.c. supply is used. The positive bias between the gate and source V_{GS} is produced in the normal way by source resistor R_3.

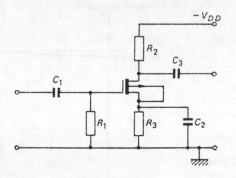

Fig. 26.11 Fet amplifier using a p-channel depletion type

27 More solid state devices

The zener diode

The zener diode is a junction diode which has a specified reverse breakdown voltage. Unlike the ordinary diode the zener is operated at its breakdown voltage. Figure 27.1 shows the characteristics for a zener diode. In the forward direction it behaves in the same way as a normal diode. When reverse biased, no current flows until the zener reaches its breakdown voltage V_Z, usually known as the zener voltage, when current suddenly begins to flow. At breakdown the voltage across the zener remains constant for a very large change of current. The zener is the semiconductor equivalent of the gas-filled diode. Zener diodes are used for shunt stabilisation and for voltage reference (see Chapter 29).

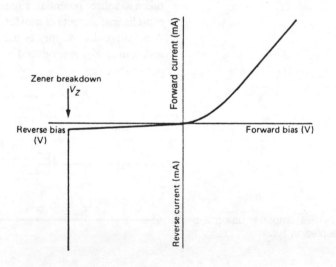

Fig. 27.1 Forward and reverse characteristics of a zener diode

The switching diode

The switching diode consists of four layers of semiconductor material as shown in Fig. 27.2. When the diode is forward biased a very small current flows until breakdown occurs (Fig. 27.3). Before breakdown

Fig. 27.2 Switching diode

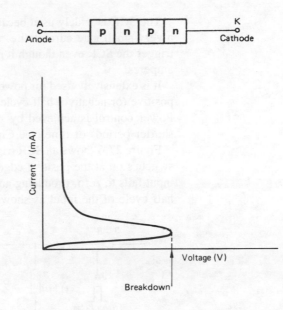

Fig. 27.3 Characteristics of a
switching diode

the diode may be considered as a switch in the 'off' position. After
breakdown it may be considered as a switch in the 'on' position.

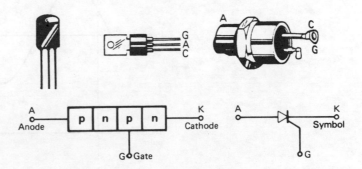

Fig. 27.4 The SCR, its symbol and
various outlines

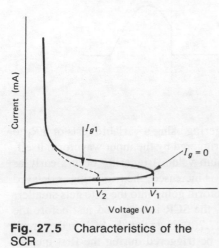

Fig. 27.5 Characteristics of the
SCR

The controlled rectifier

The controlled rectifier or the **thyristor** is another 4-layer pnpn
junction device usually made of silicon; hence the more common
name, the silicon controlled rectifier (SCR). Unlike the switching
diode the SCR has a third terminal known as the gate (Fig. 27.4).
The critical breakdown voltage may now be varied by the potential
at the gate. Figure 27.5 shows the characteristics of an SCR for two
different values of gate current. At zero gate current (i.e. a gate
potential of zero volts), the SCR has a breakdown voltage of V_1. If
the gate is given a positive voltage with respect to the cathode to drive
a gate current I_{g1}, the SCR breaks down at a lower voltage V_2. Once
the SCR starts to conduct, the gate potential has no effect on the current
through it. The SCR may be switched off only by making the anode
voltage fall below the cathode potential.

The SCR is widely used because of its fast switching and because a very small gate current (i.e. a very small power) is sufficient to trigger the SCR even though it may be handling a current of several amperes.

It is extensively used for power rectification. It turns on during the positive (or negative) half cycle only, thus producing pulsating d.c. Power control is achieved by switching on the SCR for longer or shorter periods of time (see Chapter 29).

Figure 27.6 shows an SCR triggered by a train of pulses. The SCR switches on at the positive edge of the pulse and stays on until the input falls to zero, producing an output which is part of the positive half cycle of the input as shown.

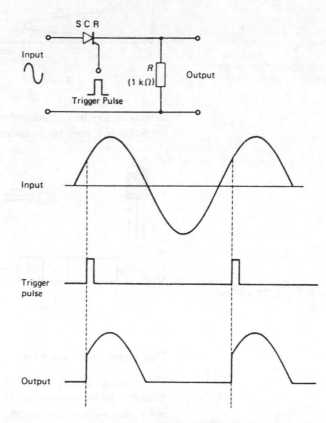

Fig. 27.6 Pulse triggered SCR

Figure 27.7 shows a.c. triggering using a variable resistor VR_1 to control the SCR. The SCR is triggered by the input waveform itself. When VR_1 is set to its minimum value, triggering occurs early as shown in Fig. 27.7(a). As VR_1 is increased, the triggering is delayed since the amplitude of the waveform going into the gate gets smaller. At the maximum value of VR_1 the SCR is triggered just before the positive peak as shown in Fig. 27.7(b). Note that the SCR in the circuit under consideration may only be triggered during the first quarter

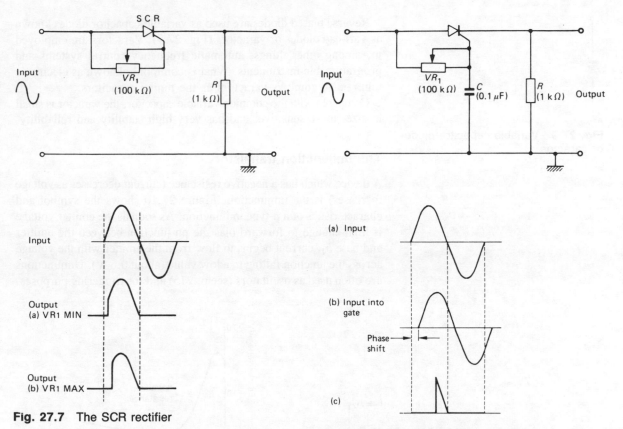

Fig. 27.7 The SCR rectifier

Fig. 27.8 SCR rectifier employing a
phase shift network VR_1-C

of the cycle, i.e. before the positive peak arrives at the gate. If it fails
to do that, it will not trigger at all and the output would be zero.

To achieve triggering during the second quarter of the cycle, i.e.
after the positive peak has passed, a phase shift network is used as
shown in Fig. 27.8. Capacitor C and resistor VR_1 act as the phase
shift network. The waveform fed into the gate suffers a delay (or a
phase shift) as shown in Fig. 27.8(b). As stated above, the SCR may
only be triggered before the positive peak arrives at the gate. But due
to the phase shift, by the time the positive peak arrives at the gate,
the positive peak of the input has already passed. Hence the SCR may
now be triggered during the second quarter of the input cycle, as shown
in Fig. 27.8(c).

The varactor

It is found that a reverse biased pn junction diode has a small
capacitance across it which varies with its reverse voltage. This fact
is used in the fabrication of integrated circuits in order to include
capacitors within the silicon slice.

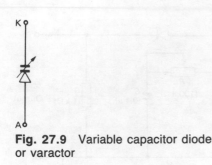

Fig. 27.9 Variable capacitor diode or varactor

Reverse biased diodes are used as variable-capacitor diodes known as vari-cap diodes or varactors (Fig. 27.9). Varactors are employed in, among other things, automatic frequency control systems and programmable instruments. What is commonly known as electronic tuning also employs varactors as the tuning capacitors.

Compared with the common variable capacitor, the varactor is small in size, more sensitive, and has very high stability and reliability.

The unijunction transistor

A device which has a negative resistance (current decreases as voltage increases) is the unijunction. Figure 27.10 shows the symbol and characteristics of a p-type unijunction. As soon as the emitter voltage is high enough to forward bias the pn junction between the emitter and base b_1, current begins to flow from the emitter with the voltage across the junction falling to a low value (about 0.6 V). Unijunctions are often used as oscillators (see p. 215) and for triggering purposes.

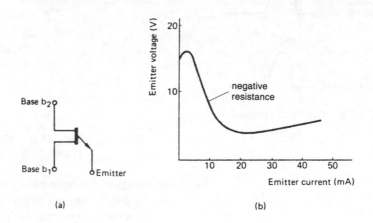

Fig. 27.10 Unijunction. (a) p-type symbol (b) characteristics

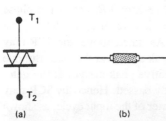

Fig. 27.11 The diac. (a) symbol (b) outline

The diac

The diac is another switching device, with two terminals T_1 and T_2, as shown in Fig. 27.11. By increasing the potential between T_1 and T_2, regardless of polarity, breakdown occurs. The diac may conduct in either direction and hence it is also known as a bidirectional diode. When breakdown occurs, the voltage across the diac falls to few volts. Breakdown voltages are in the region of 30−50 V. The diac is used as a triggering element, e.g. for SCRs.

The triac

The triac (Fig. 27.12) is another bidirectional diode with a controlling gate terminal. Breakdown occurs when the voltage between the two terminals T_1 and T_2 reaches a specified level, regardless of the polarity.

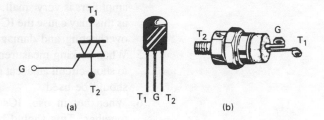

Fig. 27.12 The triac. (a) symbol
(b) outlines

The triac can be triggered into conduction by a gate signal which may be either positive or negative with respect to T_1 or T_2. Triacs are also used as triggering devices, e.g. for SCRs.

Integrated circuits

Advances in technology have resulted in improvements in reliability of electronic devices as well as a reduction in cost and size. Microelectronic circuits are simply a miniature assembly of a number of electronic components, both passive and active.

There are two types of micro-circuits: the film circuits and the monolithic integrated circuits. There are two types of film-integrated circuits, known as thick film and thin film. Both are constructed by depositing films of special resistive paste on an insulating surface. Mainly used for resistors, they may also be used for small-value capacitors and inductors.

The monolithic integrated circuit normally referred to as the integrated circuit (IC) uses a basic p-type silicon disc or chip. The silicon chip is a very thin wafer (0.02 cm thick) with a surface area (approximately 26 mm^2) equivalent to the cross-sectional area of a match stick. The chip acts as a substrate into which various components are fabricated by a process known as diffusion. Integrated circuits may contain a number of active devices, transistors, diodes, etc., as well as resistors and capacitors. *Large scale integration* (LSI) can produce ICs containing complete systems, e.g. decoders and microprocessors.

Although integrated circuits are solid devices and physically strong, electrically they are delicate and must be handled with care. The following precautions must be taken when replacing ICs:

1. ICs should be handled by their casing, avoiding touching the pins, otherwise dirt and grease may accumulate, reducing contact.
2. When soldering an IC, extreme care must be taken to ensure that it does not get too hot. Overheating will readily damage most ICs.
3. The correct specified d.c. voltages must be used. Most ICs need highly regulated d.c. supplies. This is especially important in digital applications.
4. Power dissipation of most ICs with the exception of power

amplifiers is very small. Overloading must therefore be avoided as this may cause the IC to exceed its rated dissipation, causing overheating and damage.

5. When making measurements on an IC, care must be taken not to short-circuit adjacent pins. Specially shaped or pointed probes should be used.

6. When not in use, ICs (MOS type) have their pins shorted together. This should be done whether the IC is kept on the shelf or is being packaged for transportation.

28 $R-C$ filters, clippers and clampers

A filter allows one frequency band to pass through without attenuation (weakening) and cuts off all other frequencies. The frequency at which cut-off occurs is known as the cut-off frequency f_c (Fig. 28.1).

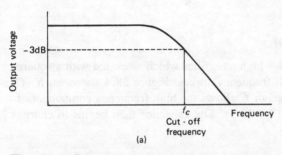

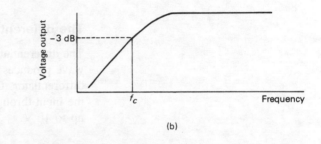

Fig. 28.1 Frequency response (a) Low pass filter (b) High-pass filter

Effect on a square wave

As explained in Chapter 3, a square wave is a complex waveform consisting of a fundamental and an infinite number of odd harmonics. A square wave can thus be thought of as consisting of a low frequency component represented by the flat top and bottom, and a high frequency component represented by the fast rising and falling edges.

When fed through a filter, the square wave suffers a distortion in its shape. In general, a low pass filter will distort the high frequency edges, rendering them less sharp and producing rounded corners as shown in Fig. 28.7(b). (A low pass filter has the same effect on a square wave as an amplifier with insufficient bandwidth.) Conversely, a high pass filter distorts the flat top and bottom of a square wave as shown in Fig. 28.5(b).

$R-C$ filters

The simplest type of a filter is an $R-C$ filter. It uses the fact that the reactance of a capacitor varies inversely with frequency, while

161

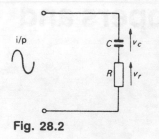

Fig. 28.2

a resistance remains constant as frequency changes. Figure 28.2 shows a capacitor in series with a resistance. At low input frequencies, the reactance of the capacitor is very large compared with resistance *R*. Hence the voltage v_c across the capacitor is large while that across the resistor v_r is low. At high input frequencies the opposite is true, namely v_c is low and v_r is high. Therefore if an output is taken across the capacitor (Fig. 28.3(b)), low frequencies will predominate while high frequencies will be greatly attenuated. In other words we have a low pass filter. Conversely, an output across the resistor produces a high pass filter (Fig. 28.3(a)). The values of *R* and *C* determine the cut-off frequency.

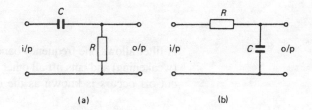

Fig. 28.3 (a) *R–C* high pass filter
(b) *R–C* low pass filter

The differentiator

The differentiator is a high pass filter which when fed with a square wave produces high frequency spikes. Figure 28.4 shows an *R–C* differentiator. Capacitor *C* allows the high frequency component of the input through (Fig. 28.5). The capacitor then begins to charge up to 10 V.

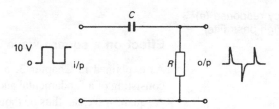

Fig. 28.4 *R–C* differentiator

Provided the time constant (the product of *C* and *R*) is short compared with the periodic time of the input waveform, the capacitor will become fully charged to 10 V well before the next high frequency component CD arrives (Fig. 28.5(a)). When the capacitor is fully charged, the voltage across the resistor, i.e. the output, is zero. The falling edge CD represents a 10 V drop and, being a high frequency component, will pass through the capacitor bringing the output voltage down to −10 V. The capacitor then begins to charge up to −10 V, and given the short time constant the capacitor will be fully charged, with the output falling to zero well before the next high frequency edge arrives and so on. The effect of a time constant that is long compared with the period of the input is shown in Fig. 28.5(b).

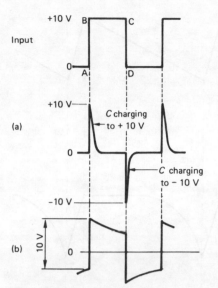

Fig. 28.5 Output from the differentiator in Fig. 28.4 (a) Short time constant (b) Long time constant

Fig. 28.6 *R–C* integrator

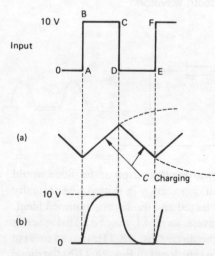

Fig. 28.7 Output from integrator in Fig. 28.6 (a) Long time constant (b) Short time constant

The integrator

The integrator is a low pass filter which when fed with a square wave produces a triangular waveform. Figure 28.6 shows an *R–C* integrator. At the first rising edge of the input (Fig. 28.7), the capacitor begins to charge up towards +10 V. Provided the time constant *CR* is long compared with the period of input, the falling edge CD arrives well before the capacitor is fully charged (Fig. 28.7(a)). The falling edge CD will attempt to charge the capacitor in the negative direction. Again, due to the long time constant, before the capacitor can charge up negatively, the next edge EF arrives, and so on. The output waveform will, therefore, be triangular in shape having a low amplitude compared with the input.

The effect of a short time constant is shown in Fig. 28.7(b). Note that for both a differentiator and an integrator, the time constant is always compared with the period of the input. A time constant of 100 μs for example is long compared with a periodic time of say 5 μs (input frequency of 200 kHz) but short compared with a periodic time of say 5 ms (an input frequency of 200 Hz).

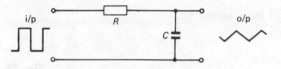

The effect on a sine wave

A sine wave is a pure waveform with no harmonics. When fed into a filter of any type it suffers no change in its shape. Its amplitude may be affected, depending on whether the frequency of the sine wave falls within the passband of the filter or outside it. In the former case, the sine wave suffers very little attenuation, while in the latter it may be very largely attenuated.

The effect on a saw-tooth waveform

The effect of integration on a saw-tooth waveform is to round off the sharp edges, as shown in Fig. 28.8. How far the sharp edges are removed is determined by the degree of integration, i.e. the time constant of the circuit. A very long time constant (a high degree of integration) will produce the effect shown in Fig. 28.8(b)).

The effect of differentiating a saw-tooth waveform is that shown in Fig. 28.9. A high degree of differentiation, i.e. a very short time constant, will produce the spiky pulse shown in Fig. 28.9(b).

Diode clipping

A **clipper**, also known as a limiter, flattens or slices the top or bottom or both of a waveform. The circuit in Fig. 28.10 clips the whole of

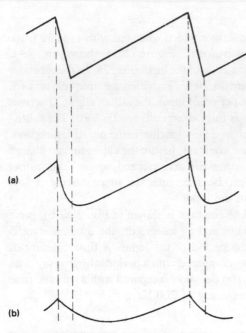

(a)

(b)

Fig. 28.8 Effect of integration on a saw-tooth waveform

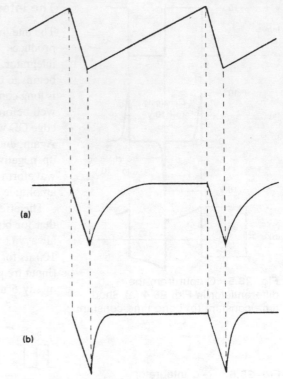

(a)

(b)

Fig. 28.9 Effects of differentiation on a saw-tooth waveform

Fig. 28.10

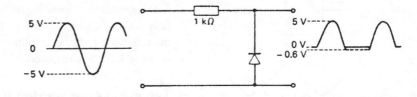

the negative half cycle. The forward voltage drop of the diode would produce a 0.6 V negative excursion shown in dotted line. Usually the forward voltage drop is neglected and diodes are assumed ideal.

To produce clipping at other levels, an e.m.f. may be added in series with the diode to forward bias the diode (Fig. 28.11(a)) or to reverse bias the diode (Fig. 28.11(b)). In the circuit in Fig. 28.11(b) the diode conducts only when its cathode is at or below −2 V, assuming ideal diode. Cathode voltages above −2 V will keep the diode reverse biased, i.e. open circuit, and hence that part of the waveform appears at the output.

A simple method of working out the output of a diode clipper with an external e.m.f. is as follows:

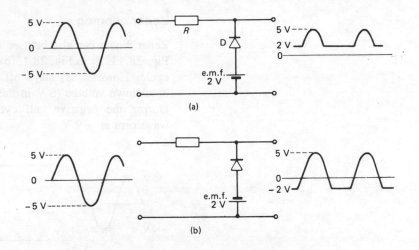

Fig. 28.11 (a) Clipping at +2 V
(b) Clipping at −2 V

1. Draw a line representing the e.m.f. level along the input waveform.
2. If the e.m.f. is connected in such a way as to forward bias the diode, then the larger part of the waveform will be clipped off and vice versa.

The circuits in Fig. 28.12 produce clipping of both positive and negative half cycles. In Fig. 28.12(a), D_1 clips the positive half while D_2 clips the negative half cycle. For ideal diodes the output would be zero. However, due to the forward voltage drop across the diodes (0.6 V for silicon diodes) clipping occurs at +0.6 V and −0.6 V.

Figure 28.12(b) shows a circuit which is in effect a doubling of that shown in Fig. 28.11(b). D_1 clips the positive half at +2 V while D_2 clips the negative half at −4 V.

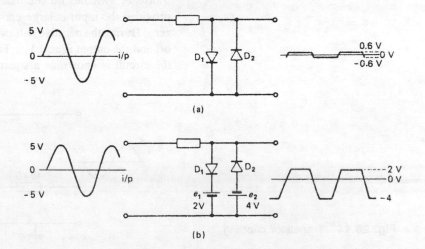

Fig. 28.12 Positive and negative clipping

Zener clipping

Zener diodes can also be used for clipping a waveform as shown in Fig. 28.13. Z_2 in Fig. 28.13(b) conducts for the whole positive half cycle. However Z_1 stays off until the input voltage exceeds its breakdown voltage (6 V in this case) clipping the input as shown. During the negative half cycle, Z_1 conducts and Z_2 clips the waveform at −9 V.

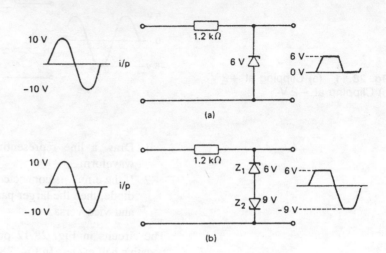

Fig. 28.13 Zener clipping

Transistor clipping

As explained in Chapter 9, clipping can be achieved by overloading an amplifier. The circuit in Fig. 28.14 produces an output which is practically a square wave. The transistor has no bias voltage, hence it is at cut-off without an input. During the positive half cycle the transistor switches on (its base going away from the emitter) and provided the input is large enough it saturates, giving an output of zero. During the negative half cycle of the input, the transistor remains off and the output is at $+V_{cc}$. The output is a square wave and hence the circuit is known as a squarer.

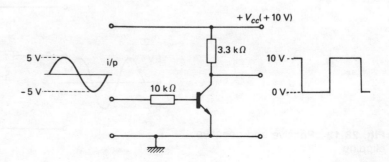

Fig. 28.14 Transistor clipping

Schmitt trigger

A very useful circuit for squaring, clipping and pulse shaping is the Schmitt trigger shown in Fig. 28.15. With no input, TR_1 base is at zero potential and the transistor is cut-off. TR_2 however is saturated since its base has a positive voltage determined by potential divider $R_2 - R_4 - R_5$. The output (TR_2 collector) is practically at zero potential. TR_2 current flowing through the common emitter resistor R_3 gives the emitters a positive voltage which reverse biases TR_1.

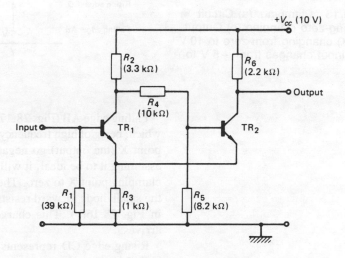

Fig. 28.15 Schmitt trigger circuit

When the input is increased in the positive direction, TR_1 begins to conduct when its base reaches TR_1 emitter voltage $+0.6$ V. When that happens, TR_1 begins to conduct, its collector potential drops and with it TR_2 base voltage. This leads to a drop in TR_2 current resulting in a drop in emitter (TR_2 and TR_1) potential causing TR_1 to conduct even more and so on. This results in TR_1 saturating and TR_2 turning off very quickly, with the output going to V_{cc}. If the input voltage is now decreased towards zero the reverse operation will occur resulting in TR_1 OFF, TR_2 ON.

The clamper

Also known as d.c. restorer, the clamper provides a.c. waveforms with a d.c. level. A clamper does not change the shape of the input signal, it just provides it with a d.c. component.

There are many occasions where it is important to retain the d.c. level of a signal. The d.c. level is sometimes lost due to a.c. coupling and has to be restored, e.g. a pulse train following differentiation or a video signal following a coupling capacitor. Clamping is also used to provide Class C biasing for transistors when used in oscillators for example (see Chapter 30). Figure 28.16 shows a clamper circuit.

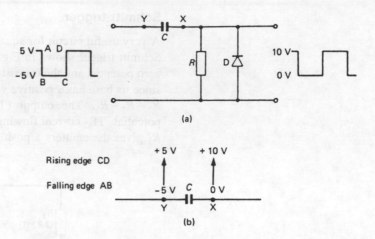

Fig. 28.16 Clamping. (a) Circuit producing zero clamping (b) Output (point X) changing from zero to 10 V as the input changes from −5 V to +5 V

Falling edge AB (Fig. 28.17) of the input represents a drop of 10 V which, being of high frequency, will pass through the capacitor making point X (the output) go negative. The diode is forward biased and assuming it to be ideal, it will act as a short circuit across the output clamping point X to zero. The capacitor quickly charges up through the small diode forward resistance in the negative direction as shown in Fig. 28.16(b). This charge is retained till the next rising edge arrives.

Rising edge CD represents a 10 V positive step which, being of high frequency, passes through the capacitor shifting point X from zero volts to +10 V (Fig. 28.16(b)). The diode is now reverse biased (i.e. open circuit) with 10 V across resistor *R*. Current begins to flow

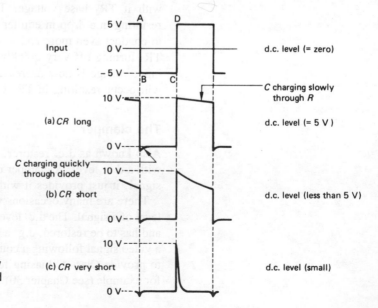

Fig. 28.17 Output of clamper in Fig. 28.16 (a) with time constant very long, (b) with time constant short and (c) with time constant very short

and attempts to charge the capacitor in the opposite direction; the voltage at the output begins to fall. However, if the time constant *CR* is long compared with the periodic time of the input, the capacitor is unable to lose its negative charge and the output remains at practically the same potential till the next falling edge arrives (Fig. 28.17(a)).

For a time constant that is not long compared with the period of the input, the output is as shown in Fig. 28.17(b). With a short time constant the circuit becomes a differentiator, having its negative-going spikes fully clipped (Fig. 28.17(c)).

To ensure a long time constant, resistor *R* may be removed, making the time constant equal to *C* × *r* where *r* is the reverse resistance of the diode.

The diode may be reversed to produce zero clamping with negative-going waveform (Fig. 28.18). Note here that the input itself has a d.c. level which is removed by the capacitor and hence does not affect the output.

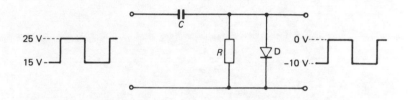

Fig. 28.18 Zero clamping with negative going output

To draw the output waveform for a given clamping circuit, first redraw the input waveform and if the diode is connected in such a way as to clip positive excursions, the waveform will be negative-going only and vice versa, having exactly the same amplitude as the input.

Clamping circuits can be used to clamp waveforms other than square waves. Figure 28.19 shows a circuit for sine wave clamping.

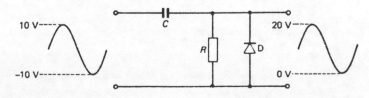

Fig. 28.19 Clamping of a sine wave

Clamping at levels other than zero are possible by adding an e.m.f. in series with the diode to give it a forward bias, as in Fig. 28.20(a), or a negative bias as in Fig. 28.20(b). In Fig. 28.20(b) the diode is given a reverse bias of 5 V thus preventing the output from going above +5 V. Since a clamping retains the amplitude of the input, the output drops to −15 V resulting in a total amplitude of 20 V.

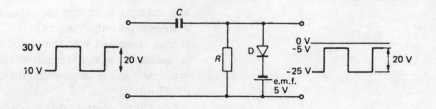

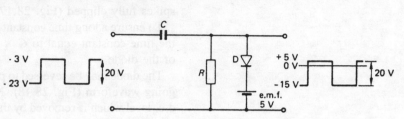

(a)

(b)

Fig. 28.20 (a) Clamping at −5 V (b) Clamping at +5 V

29 Power supplies II

Rectification

There are two types of rectifier.

1. **Half-wave rectifier**, shown in Fig. 29.1. Diode D_1 in Fig. 29.1(a) allows only the positive half cycles through, producing a positive-going output. When the diode is reversed as in Fig. 29.1(b) negative-going half cycles are produced. The output waveform contains a d.c. level which is 0.318 (approximately one-third) of the peak voltage (Fig. 29.2).

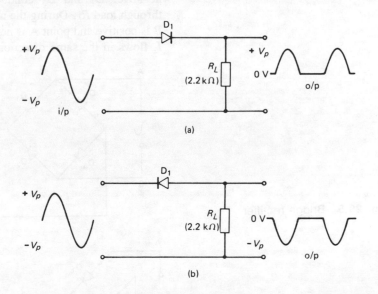

(a)

(b)

Fig. 29.1 Half-wave rectifier
(a) Positive-going (b) Negative-going

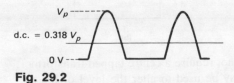

Fig. 29.2

2. **Full-wave rectifier**, shown in Fig. 29.3. A transformer with a centre tapped secondary is used. Two equal and opposite waveforms are developed across each half of the secondary winding. For one half of the input, point A is positive and point B is negative with respect to the centre tap, and vice versa for the other half cycle. When A is positive, D_1 conducts and current I_1 flows as shown. When B is positive, D_2 conducts and current I_2 flows through R_L in the same

171

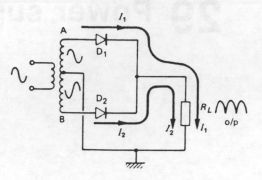

Fig. 29.3 Full-wave rectifier using a centre tap transformer

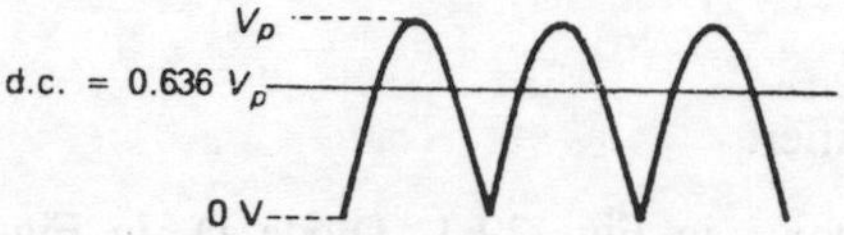

Fig. 29.4 D.C. level of a full-wave rectified sine wave is 0.636 of the peak

direction as I_1. The output waveform is therefore as shown in Fig. 29.4 with a d.c. level double that of the half-wave rectifier at 0.636 (approximately two-thirds) of the peak voltage.

The bridge rectifier

Another circuit that produces full-wave rectification is that shown in Fig. 29.5 and known as the bridge rectifier. During the positive half cycle of the input (Fig. 29.6(a)), point A is positive and point B is negative. D_1 and D_3 conduct and current I_1 flows downwards through load R_L. During the negative half cycle (Fig. 29.6(b)) point B is positive and point A is negative. D_2 and D_4 conduct and current I_2 flows in the same direction through R_L.

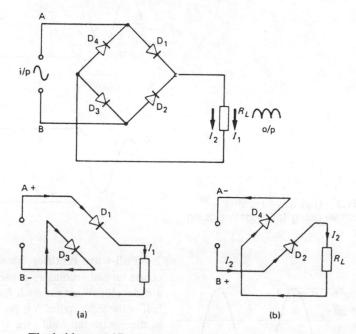

Fig. 29.5 Bridge rectifier

Fig. 29.6

(a)   (b)

The bridge rectifier does not require a centre tapped transformer. However, a transformer may be used to alter the level of the a.c. input to the rectifier.

The reservoir capacitor

In order to reduce the a.c. variation of the rectified waveform, a reservoir capacitor C_1 is used as shown in Fig. 29.7. C_1 charges up to the peak voltage and then discharges through the load R_L,

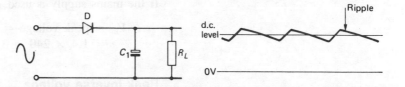

Fig. 29.7 D.C. power supply with reservoir capacitor C

preventing the voltage from falling rapidly. Figure 29.8 shows the effect of the reservoir capacitor on a half-wave and a full-wave rectifier. In both cases the output contains a higher d.c. voltage with a superimposed small a.c. ripple. The amplitude of the ripple is determined by the time constant CR of the reservoir capacitor and the load resistance. Reservoir capacitors therefore have large capacitances of between 100 and 5000 μF.

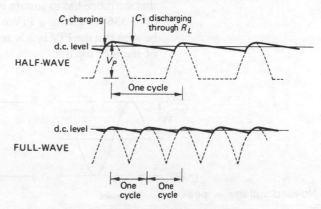

Fig. 29.8 The effect of the reservoir capacitor on a rectified sine wave

A comparison between the two waveforms in Fig. 29.8 shows that full-wave rectification has the following advantages:

1. The reservoir discharge time is shorter, giving a smaller a.c. ripple.
2. The fundamental frequency of the ripple is twice that of the a.c. supply, whereas for half-wave, the ripple frequency is equal to the supply frequency. For example, using the mains as the a.c. input, the ripple produced by full-wave rectification has a frequency $2 \times 50 = 100$ Hz as compared with 50 Hz for half-wave rectification. Full-wave ripple is easier to filter out due to its higher frequency, as will be seen later.

No-load voltage

The no-load voltage is the terminal voltage of the power unit when load current is zero, i.e. when the load is removed. Figure 29.9 shows a simple power supply with no load resistance. Reservoir capacitor C charges up to the peak voltage as normal. However with the absence of a load resistor, the capacitor retains the charge across it, thus producing a steady output (equal to the peak voltage) without any ripple. The no-load voltage is, therefore, the highest terminal voltage. If the mains supply is used, the no-load terminal voltage

$$V_{nl} = \text{peak voltage} = 1.4 \times \text{r.m.s. voltage}$$
$$= 1.4 \times 240 = 336 \text{ V}$$

Peak inverse voltage

An important factor that must be considered in choosing the diodes for d.c. power supplies is the maximum voltage the device has to sustain across it during the non-conducting half cycle known as the peak inverse voltage (PIV). Consider the circuit in Fig. 29.9. The highest potential at the cathode of diode D is the no-load voltage 336 V. The anode potential varies between the positive peak of $+336$ V and a negative peak of -336 V. The maximum peak voltage that the diode has to sustain is when the anode is at the negative peak of -336 V, giving a PIV of $336+336 = 672$ V. It can therefore be seen that the PIV is $2 \times$ no-load voltage, i.e. $2 \times$ the peak voltage of the a.c. input.

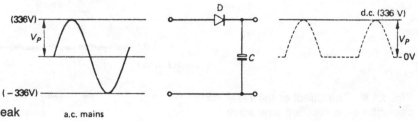

Fig. 29.9 No-load voltage = peak input voltage

R–C smoothing

The a.c. ripple may be reduced using a low pass filter or smoothing circuit. R_1 and C_2 in Fig. 29.10 form a simple $R-C$ filter. To provide adequate ripple attenuation, the time constant $R_1 C_2$ must be very large compared with the periodic time of the ripple waveform.

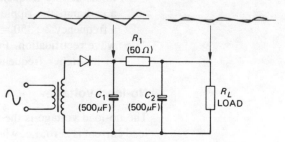

Fig. 29.10 D.C. power supply with $R-C$ filter

For a given time constant, the shorter the periodic time of the ripple (i.e. the higher its frequency), the more effective is the smoothing. Hence it is easier to smooth out full-wave ripple than half-wave ripple.

Smoothing capacitor C_2 is large, having a value comparable to that of the reservoir capacitor C_1 of between 100 to 5000 μF. Smoothing resistor R_1 however is given a small value, otherwise a large d.c. drop will develop across it reducing the terminal voltage. Values of smoothing resistors are in the region of 1 to 100 Ω depending on the load current.

L–C smoothing

A more effective smoothing circuit is shown in Fig. 29.11. L_1 and C_2 form a low pass filter. Smoothing choke L_1 has a large inductance (100 mH–10 H) which tends to smooth out variations in current through it, reducing further the a.c. ripple at the output. Its low resistance is a further advantage while its size is a disadvantage. Note that while the reservoir capacitor C_1 increases the d.c. level of the power unit, the smoothing circuit leaves the d.c. level virtually unchanged. It only attenuates the ripple content.

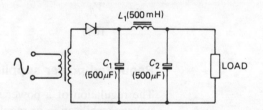

Fig. 29.11 D.C. power supply with L–C filter L_1–C_2

Regulation

An increase in the current supplied by a power unit causes the terminal voltage to drop. This is due to its internal resistance which is the sum total of the resistance of the transformer winding, diodes, smoothing resistor or choke. Figure 29.12 shows the change in the terminal voltage of a d.c. supply as load current is increased. This curve is known as the regulation curve. The terminal voltage is at a maximum when the load current is zero, i.e. at no load. When it is supplying the specified full load current, the terminal voltage is referred to as the full load voltage.

Regulation is given as a percentage and it is defined as

$$\frac{\text{No load voltage} - \text{Full load voltage}}{\text{Full load voltage}} \times 100\%$$

$$= \frac{V_{nl} - V_{fl}}{V_{fl}} \times 100\%$$

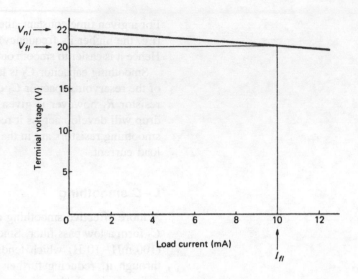

Fig. 29.12 Regulation curve for an unstabilised power supply

For example if a d.c. supply provides 22 V when on no load and 20 V when supplying a full load current of 10 mA, its regulation is

$$\frac{V_{nl} - V_{fl}}{V_{fl}} \times 100\% = \frac{22 - 20}{20} \times 100\% = \frac{2}{20} \times 100\%$$

$$= 10\%$$

Regulated power supplies

The regulation of a power supply can be improved by the use of voltage stabilisation or regulation. The terminal voltage may in this way be maintained at a constant level as the load current is varied. There are two types of stabilisation: the shunt and the series.

Shunt regulators

Figure 29.13 shows a block diagram for a shunt regulator. Figure 29.14 shows a circuit for a shunt regulated power supply in which the stabiliser element is the zener diode. The zener operates at its

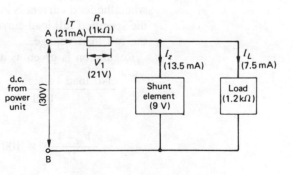

Fig. 29.13 Block diagram for a shunt regulator

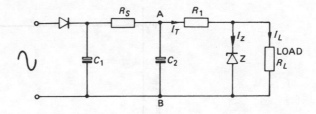

Fig. 29.14 A shunt regulated power supply

breakdown voltage, keeping the voltage across it and hence the output voltage constant for a very large variation of current through it.

Shunt regulation is based on a current-sharing principle whereby the sum of the load current I_L and the zener current I_z is kept constant. If the load current is increased by say 2 mA, then the shunt stabiliser current is decreased by the same amount of 2 mA and vice versa.

The series dropper resistor R_1 takes the total current, thus developing a voltage V_1 across it equal to the difference between the unregulated d.c. voltage V_{AB} and the zener breakdown voltage V_z.

$$V_1 = V_{AB} - V_z$$

Using the values given in Fig. 29.13 $V_0 = V_z = 9$ V.

Voltage across R_1, $V_1 = V_{AB} - V_z = 30 - 9 = 21$ V.

Therefore, total current $I_T = \dfrac{21 \text{ V}}{1 \text{ k}\Omega} = 21$ mA

Load current $I_L = \dfrac{\text{Voltage across load}}{\text{Load resistance}} = \dfrac{9 \text{ V}}{1.2 \text{ k}\Omega}$

$$= 7.5 \text{ mA}$$

Therefore

Zener current $I_z = I_T - I_L = 21 - 7.5 = 13.5$ mA

If the load current is now reduced to 5 mA, a drop of $7.5 - 5 = 2.5$ mA, the current taken by the zener will increase by 2.5 mA to a new value of $13.5 + 2.5 = 16$ mA.

Under no-load condition when $I_L = 0$, the whole of the total current I_T will now be taken by the zener: $I_T = I_z$. This is one disadvantage of the shunt regulator, namely that regardless of the load, the power supply has to provide a constant maximum current of I_T.

Figure 29.15 shows a typical regulation curve for a shunt stabilised power supply, for the circuit given in Fig. 29.14. The load voltage begins to fall rapidly as the load current increases above the full load current of just under 21 mA. At this value the load current is almost equal to the total current I_T. The fall in the voltage occurs when the zener current is too small for breakdown to occur. For good regulation, the unregulated d.c. voltage is normally chosen to be three times the breakdown voltage of the zener.

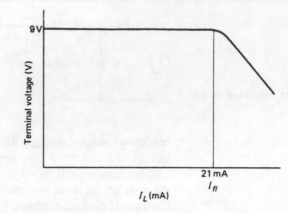

Fig. 29.15 Regulation curve for a stabilised power supply

Fig. 29.16 Block diagram for a series stabilised power supply

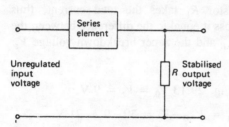

Series regulation

A better and more efficient stabiliser is the series stabiliser, which employs a transistor or a thyristor as the series element. Figure 29.16 shows a simple block diagram for a series stabiliser. It consists of a series control element with a bleed resistor R to provide a minimum load current.

Series transistor stabiliser

A basic circuit for a series stabiliser employing a transistor is shown in Fig. 29.17. The output is taken at the emitter of the TR_1 and, as can be seen from Fig. 29.18 where the same circuit is redrawn, TR_1 is used as an emitter follower. The zener maintains the base of the transistor at a constant potential. The emitter follows the base potential so that it is 0.6 V (for silicon) below the base voltage for a normal forward bias, thus keeping the output voltage constant.

The emitter follower acts as a current amplifier which helps to supply large load currents for a small input current. The zener acting as a regulator as well as a voltage reference diode takes a smaller current compared with the zener used in the shunt stabiliser. For good regulation the zener current should be approximately 5 times the base current.

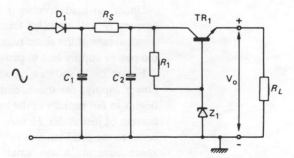

Fig. 29.17 A series stabilised power supply

The simple series stabiliser considered above suffers from two main disadvantages:

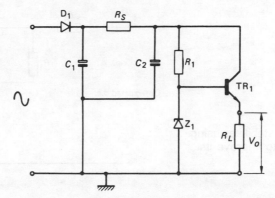

Fig. 29.18 Circuit in Fig. 29.17 redrawn to show the emitter follower configuration of TR_1

1. For large load currents, power zeners and high gain transistors are required.
2. Limited sensitivity.

The first may be overcome by increasing the current gain by means of an additional transistor TR_2 forming a second emitter follower stage as shown in Fig. 29.19. Load current may be very large (in amps) while zener current remains very small. Sensitivity, i.e. voltage stability, may be improved by amplifying the output voltage before

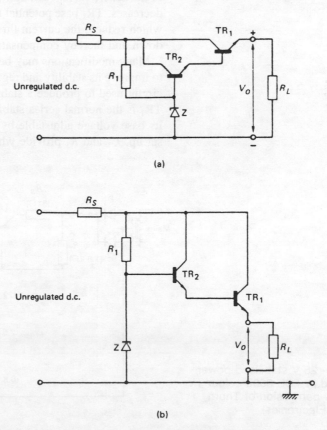

(a)

(b)

Fig. 29.19 Series stabilised power supply using two emitter followers (the circuit is redrawn in (b))

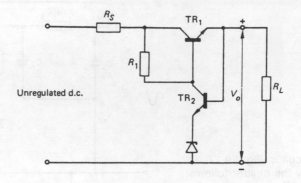

Fig. 29.20 Series stabiliser using error amplifier TR_2 to improve its sensitivity

comparing it with the reference zener voltage as shown in Fig. 29.20. TR_1 is the normal series transistor, while TR_2 is the error voltage amplifier. The zener is a reference diode only and hence may have small power rating.

TR_2 compares the output voltage with the reference voltage of the zener. Change in the output voltage is amplified and fed into the base of TR_1 which maintains the output voltage constant. For example, if due to some external disturbance the output voltage V_o increases, TR_2 base goes up away from the emitter which is at the fixed voltage of the zener, current through TR_2 increases and its collector voltage decreases. TR_1 base potential thus suffers a drop towards the emitter which reduces the current through TR_1, making load voltage V_o go down and thereby compensating for the original increase.

Many modifications may be made to the series stabiliser in order to improve its stability and regulation. Figure 29.21 shows a practical circuit used to produce a stabilised 25 V for a colour TV receiver. TR_1 is the normal series stabiliser. TR_2 is a voltage amplifier with its base voltage adjustable by VR_1 to enable the 25 V output to be set up. C_1 and R_3 provide what is known as electronic smoothing.

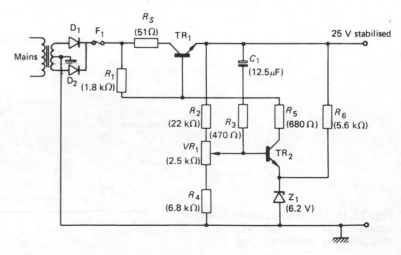

Fig. 29.21 25 V stabilised power supply used in TCE 8000 colour TV chassis (By permission of Thorn Consumer Electronics)

The 100 Hz ripple on the output is fed through TR_2 back to the base of TR_1 in anti-phase to the ripple present there, thus cancelling it out.

Overload protection circuit

One of the problems encountered by a series stabiliser is the protection of the series transistor. An excessive current through the transistor due to an overload or a short circuit may result in permanent damage. One method of protection is shown in Fig. 29.22. TR_2 is the overload protection transistor. R_1 falling in the return path of load current I_L develops a p.d. across it which provides the transistor with a b–e voltage. Provided the load current is normal, the voltage across R_1 is small, keeping TR_2 off. If the load current is increased above its normal value, the potential across R_1 increases giving a forward bias to the transistor, and TR_2 begins to conduct. When TR_2 conducts, it takes current away from TR_1 thus giving it protection. A trip switch may be incorporated to disconnect the power supply if load current increases above its safe value.

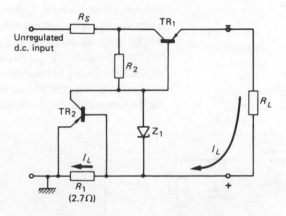

Fig. 29.22 Series stabilised power supply employing overload protection transistor TR_2

Inverters

An inverter produces sine wave signals from a d.c. input. They often include arrangements for output voltage regulation. They are mainly employed as back up generators in cases of power failure.

Sine wave inverters may be designed around class A or B oscillators. However, because of the high losses associated with such linear operations, switching elements are used to produce square waves which are then filtered to obtain a sine wave output (Fig. 29.23).

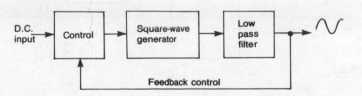

Fig. 29.23

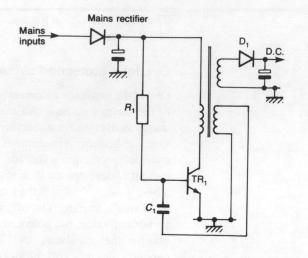

Fig. 29.24

Converters

Converters change a d.c. input voltage into another d.c. level. It consists of an inverter followed by a rectifier. Figure 29.24 shows a simple converter arrangement employing a blocking oscillator. The output from the oscillator is a square wave with its space determined by time constant R_1C_1. Rectifier diode D_1 is connected across a secondary winding to produce a d.c. output. A more efficient converter is shown in Fig. 29.25, in which TR_1 and TR_2 are two blocking oscillators which drive current into the transformer in turn.

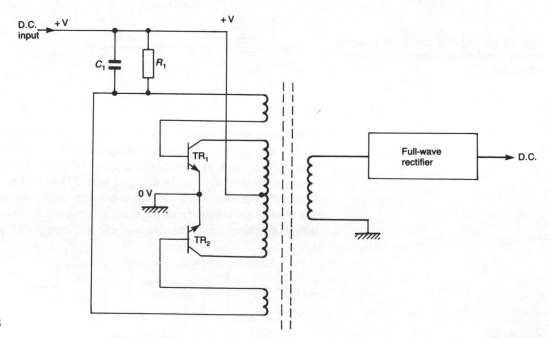

Fig. 29.25

Switched mode power supplies (SMPS)

A more efficient power supply is the switched mode. In this type the series regulating element is a switching element (SCR or transistor) which is turned on and off by a pulse on an a.c. waveform to adjust and regulate the d.c. output.

The switching mode power supply is in essence a converter. It converts unregulated d.c. into a switched or pulsating d.c. and back again into regulated d.c. (Fig. 29.26). The switching speed determines the a.c. or ripple frequency at the output which is removed by a low-pass smoothing circuit.

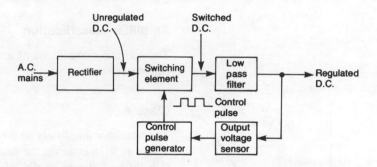

Fig. 29.26

In Fig. 29.26, mains a.c. is first rectified to produce an unregulated d.c. which goes into the anode of the switching element. The switching element, which may be a transistor or a thyristor, is opened and closed at regular intervals by a pulse from the control unit. It is used to charge up a reservoir capacitor. The charge across the capacitor is determined by the period during which the switch is closed. Regulation is obtained by making the time intervals when the switching element is open and closed (i.e. the mark-to-space ratio of the control pulse) dependent on the d.c. output through the output voltage sensor. A drop in the output voltage is thus compensated for by a wider pulse closing the switching element for a longer period, and vice versa.

30 Amplifiers

Amplifier classification

The operating conditions of amplifiers may be classified into three categories.

Class A

The transistor amplifiers so far considered have been biased in such a way as to operate on the linear part of their characteristics. The Q-point was chosen in the middle to give a maximum undistorted output. These amplifiers are said to operate under class A condition. Figure 30.1 shows a transistor transfer characteristic. Point A represents class A biasing. The input signal is small enough to keep the b−e junction forward biased. Hence the transistor conducts for the complete cycle, i.e. for 360°.

Class A amplifiers have the advantage of producing an undistorted output, hence their wide use as pre-amplifiers and drivers, and for i.f. and r.f. stages. However, they suffer from low efficiency (about 30%) due to the fact that the transistor conducts and hence dissipates power regardless of the presence or the absence of a signal input.

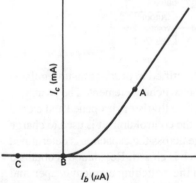

Fig. 30.1 Points A, B and C on the transfer characteristics representing the operating points for classes A, B and C respectively

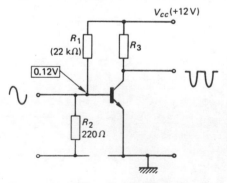

Fig. 30.2 Class B amplifier: R_1-R_2 provides a small forward bias V_{be} (0.12 V) to avoid non-linearity

Class B

Here the transistor is biased at cut off. Point B in Fig. 30.1 shows class B biasing for a transistor. The transistor conducts for one half cycle only (i.e. 180°) as shown in Fig. 30.2. Higher efficiency is possible under class B (between 50 to 60%) since the transistor dissipation is limited only to one half of the input cycle. Class B amplifiers are used in push−pull power stages and often have a small forward bias to reduce distortion.

Class C

The transistor is now biased beyond cut off (point C in Fig. 30.1). The transistor conducts for less than half a cycle (i.e. less than 180°).

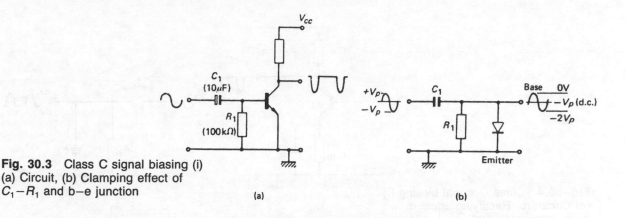

Fig. 30.3 Class C signal biasing (i)
(a) Circuit, (b) Clamping effect of
$C_1 - R_1$ and b–e junction

(a)

(b)

The output is therefore in the form of a pulsating waveform as shown
in Fig. 30.3. It has very high efficiency of between 65 to 85%. Class
C amplifiers are used in oscillators and r.f. power amplifiers.

Biasing

Class A biasing is provided by a potential divider as explained in
previous chapters, giving the b–e junction the necessary forward bias.

Class B amplifiers are biased at cut off, i.e. at a base–emitter
voltage of zero. No bias chain is therefore necessary. However, to
avoid operating in the non-linear part of the characteristics, the
transistor is given a small forward bias of 0.1 to 0.2 V by potential
divider $R_1 - R_2$ shown in Fig. 30.2. The Q-point is now slightly
above cut off.

Class C amplifiers are biased beyond cut off. In other words the
b–e junction is given a reverse bias. This is achieved by using the
input signal to bias the transistor. Hence class C biasing is also known
as *signal biasing*. Two methods may be used to achieve class C
biasing.

The most effective method is that shown in Fig. 30.3(a). Without
a signal input the base is at zero potential. As shown in Fig. 30.3(b),
the b–e junction together with C_1 and R_1 act as a clamper for
incoming signals, reproducing the input waveform together with a
negative d.c. of approximately peak voltage $-v_p$. The base is
therefore given a reverse bias approximately equal to the peak voltage
of the input. The amount of the reverse bias (or the degree of class
C biasing) may be reduced by reducing the time constant $C_1 R_1$
(usually by choosing a smaller value of R_1).

A second method is that shown in Fig. 30.4. Here the emitter is
given a positive potential by the charge on C_3. With the base at zero
potential, a positive emitter provides a reverse bias to the b–e junction.
As shown in Fig. 30.4(b), for the input signal the b–e junction
together with C_3 and R_3 act as a rectifier giving C_3 a positive charge.

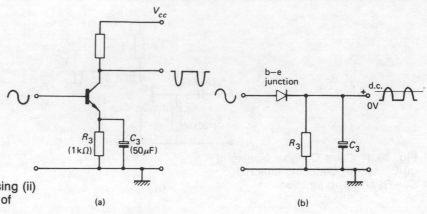

Fig. 30.4 Class C signal biasing (ii) (a) Circuit (b) Rectifying effect of b−e junction R_3 and C_3

Cascade amplifiers

Figure 30.5 shows a 2-stage $R-C$ coupled a.f. transistor amplifier. TR_1 and TR_2 are biased in class A by R_1-R_2 and R_5-R_6 respectively. The two stages are isolated from each other as far as d.c. is concerned by coupling capacitor C_2.

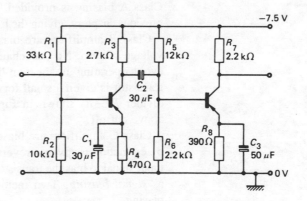

Fig. 30.5 Two-stage a.f. amplifier connected in cascade

Bandwidth

A typical frequency response curve of an amplifier is shown in Fig. 30.6. It shows that the output or gain of the amplifier remains constant at the mid-frequency range but falls at high and low frequencies.

The drop at the low-frequency end is due to the coupling capacitor C_2 in Fig. 30.5. As the frequency falls, the reactance of the coupling capacitor increases, reducing the amplitude of the signal fed into TR_2. Decoupling capacitors C_1 and C_3 also reduce the low-frequency response but their effect is very small and may be neglected.

The fall at the high-frequency end is due to what is known as *stray* or *interelectrode capacitors* across the transistor. Stray capacitors exist between the various electrodes of the transistor. They have the effect of shunting the input and output thus reducing the gain of the amplifier at high frequencies. Every transistor has what is known as a cut-off

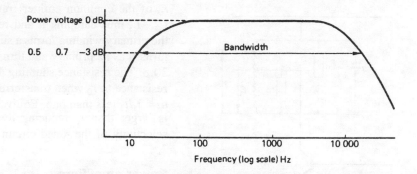

Fig. 30.6 Typical a.f. frequency response

frequency at which its current gain is too low for amplification to take place. This is an important restriction on the use of a transistor at high frequencies.

The bandwidth of the amplifier is given by the 3 dB points where the output voltage falls to 70% of its maximum value (or where the output power is half its maximum value).

I.F. amplifiers

The i.f. amplifier is a voltage amplifier using a tuned circuit as the load. It operates at intermediate frequencies such as 470 kHz for a.m. radio, 10.7 MHz for f.m. radio, and 39.5 MHz for a TV receiver.

Figure 30.7 shows a typical i.f. amplifier used in an a.m. radio receiver. $R_1 - R_2$ form a bias chain for TR_1, C_2 is the bias decoupling capacitor, C_4 is the emitter decoupling capacitor, and R_3 the emitter resistor for d.c. stabilisation. $C_1 - L_1$ and $C_3 - L_3$ are resonant circuits tuned to an i.f. of 470 kHz. Transformer coupling is used for both the input and output. The purpose of the tapping on the primary of the output transformer T_2 is to improve the selectivity of the i.f. amplifier. Without a tapped primary the low output resistance

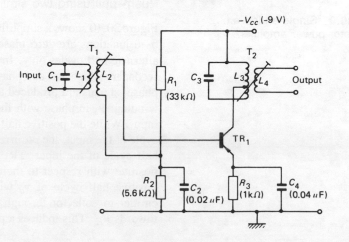

Fig. 30.7 I.F. amplifier suitable for use in a.m. radio receivers

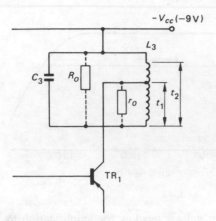

Fig. 30.8 The effect of tapping a tuned circuit. L_3 is effectively an auto-transformer

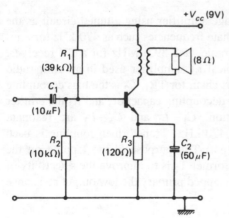

Fig. 30.9 Single-ended a.f. transistor power amplifier

r_0 of the common emitter transistor falls across the tuned circuit $C_3 - L_3$ thus damping it and reducing its selectivity. With tapping, the primary winding forms a single winding step-up transformer with turns t_1 as its primary and turns t_2 as the secondary as shown in Fig. 30.8. The resistance shunting the tuned circuit is now the equivalent resistance to r_0 when transferred to the secondary, with a turns ratio $n = t_1/t_2$ less than one. Equivalent resistance $R_0 = r_1/n^2$ (see p. 50) is larger than r_0 reducing the damping effect and improving the selectivity of the tuned circuit.

Power amplifiers

Amplifiers so far considered are what have been referred to as voltage amplifiers, the purpose of which is to produce a large voltage swing. Where output power is required, for instance to feed into a loudspeaker or an r.f. aerial or to drive a motor, power amplifiers must be used. They have high power gain which is achieved by high voltage and current gains.

Class A operation

Figure 30.9 shows a basic single-ended transistor output stage. Class A operation must be used for undistorted output. Efficiency is very low with a large drain from the d.c. supply. It can deliver small power outputs only. It may be used in car radios where current drain is unimportant.

Push–pull operation

Push–pull output stages are almost universally used in modern transistorised amplifiers. Push–pull amplifiers use two transistors operating in class B, each one amplifying one half cycle of the input.

Push–pull using two similar transistors

Figure 30.10 shows a simplified circuit of a push–pull amplifier. The b–e junctions are zero biased with each transistor conducting for alternate half cycles only. Input transformer T_1 has a centre tapped secondary which acts as a phase splitter. Two equal and opposite (anti-phase) signals are produced across each half cycle of the secondary winding: v_a in phase with the input and v_b in the anti-phase to the input. While the positive half cycle of v_a represents the positive half cycle of the input, the *positive* half cycle of v_b represents the *negative* half cycle of the input. TR_1 and TR_2 conduct when their bases go positive with respect to their emitters. Thus TR_1 conducts for the positive half cycle of v_a with current i_1 flowing as shown from emitter to collector through the top half of output transformer T_2 towards V_{cc}. This induces a positive half cycle across the secondary

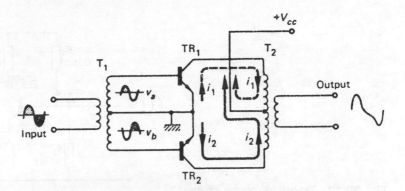

Fig. 30.10 Push–pull amplifier using two identical transistors with a transformer phase splitter

of the T_2. TR_2 conducts for the positive half cycle of v_b with its current i_2 flowing upwards (opposite to i_1) through the bottom half of transformer T_2 inducing a negative half cycle at the secondary winding. The centre tapped output transformer therefore combines the two half cycles to produce one complete cycle. TR_1 and TR_2 are connected in the common emitter configuration thus having a relatively high output impedance. Since the load to the output stage is usually very small, e.g. a 10 Ω loudspeaker, a matching transformer T_2 is always used.

The output waveform produced by the zero biased push–pull amplifier suffers from the cross-over distortion shown in Fig. 30.10. This is due to the non-linear part of the characteristics of the two transistors. It takes place during the period when one transistor begins to switch off and the other transistor begins to switch on. To overcome cross-over distortion, the transistors are given a small forward bias of between 0.1 and 0.2 V, as shown in Fig. 30.11, where $R_1 - R_2$ provide a bias chain for both transistors. The non-linearity of the two transistors cancel each other producing an undistorted output.

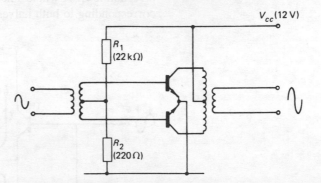

Fig. 30.11 $R_1 - R_2$ bias chain providing small forward bias to remove cross-over distortion

Transistor phase splitters

Figure 30.12 shows a phase splitter using an npn transistor. R_3 and R_4 have equal value to give two equal and opposite sine waves at the two outputs taken at the emitter and the collector. For maximum

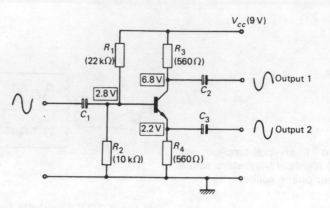

Fig. 30.12 Transistor phase splitter

undistorted outputs, the ratio of $R_1:R_2$ should be between 2:1 and 3:1. Typical quiescent d.c. voltages are shown on the circuit.

The complementary push–pull

The complementary push–pull power amplifier avoids the use of either a phase splitter at the input or a transformer at the output. It uses two symmetrical transistors, a pnp and an npn, known as a complementary pair. It uses the fact that a positive-going signal switches an npn transistor on, while a negative-going signal switches a pnp transistor on. Figure 30.13 shows the basic circuit for the complementary push–pull. TR$_1$ and TR$_2$ are biased in class B (i.e. at cut off). Two d.c. supplies are used: $+V_{cc}$ and $-V_{cc}$. The positive half cycle of the input switches TR$_1$ on and TR$_2$ off. TR$_1$ current i_1 flows towards $+V_{cc}$, upwards through load resistor R as shown. For the negative half cycle, TR$_2$ conducts with its current i_2 flowing away from $-V_{cc}$ towards chassis downwards through load R in opposite direction to i_1. A whole sine wave will thus appear across the load, corresponding to both halves of the input cycle. It should be noted

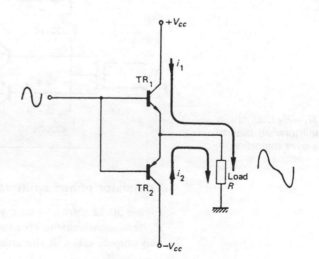

Fig. 30.13 Basic circuit for push–pull complementary amplifier

that since the output is taken at the emitters of the transistors, TR_1 and TR_2 are therefore common collector or emitter follower amplifiers.

Figure 30.14 shows a complete circuit for a complementary push–pull power amplifier together with the driver. The circuit is now modified to use one single d.c. supply. TR_1 is the driver amplifier biased in class A by resistor chain $R_1 - R_2$. When the d.c. supply is switched on, TR_1 conducts normally. Coupling capacitor C_3 has no charge across it. Hence point A, the emitter of TR_2 and TR_3, is at zero potential. The bases of TR_1 and TR_2 however are at a positive potential determined by the voltage at the collector of TR_1. This positive voltage switches TR_2 on. TR_3 being a pnp transistor is switched off. TR_2 conducts as shown with its current i_2 charging C_3.

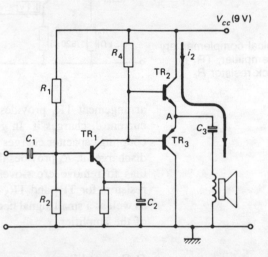

Fig. 30.14 Complementary push–pull amplifier using independent biasing chain for driver TR_1

As C_3 charges up, the voltage at point A rises. This continues until TR_2 switches off which takes place as soon as its emitter voltage (point A) is the same potential as its base. If TR_1 is designed to have a collector voltage of $\frac{1}{2} V_{cc}$ (4.5 V), then TR_2 switches off as soon as point A is at $\frac{1}{2} V_{cc}$ potential. The circuit remains balanced in this way with a $\frac{1}{2} V_{cc}$ applied across each transistor. TR_2 and TR_3 are biased at cut off (class B) with a b–e junction bias of zero volts.

When an input signal is applied, TR_1 conducts for the whole cycle, amplifying the input and providing the necessary drive for the output transistors TR_2 and TR_3. The complementary pair then amplify the signal as described for the basic circuit above.

The circuit in Fig. 30.14 suffers from d.c. instability. Any change in TR_1 current produces a change in the quiescent condition of the output pair which may result in distortion. To avoid this, negative d.c. feedback is used to bias TR_1 as shown in Fig. 30.15. The d.c. voltage at point A ($\frac{1}{2} V_{cc}$) is fed back to the base of TR_1 via resistor R_F. In this circuit the loudspeaker load is connected to the d.c. supply with C_3 as the coupling capacitor. It should be noted that with this

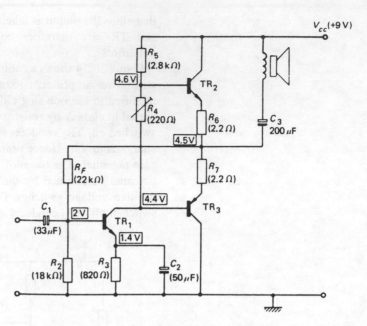

Fig. 30.15 A typical complementary push–pull power amplifier. TR_1 is biased via feedback resistor R_F

arrangement TR_3 provides the charging current for C_3 while TR_2 current discharges it. In general the transistor 'in series' with the coupling capacitor charges the latter while the transistor 'in parallel' discharges it. R_4 provides the output transistors with a small forward bias to remove cross-over distortion. R_6 and R_7 are the emitter resistors for TR_2 and TR_3 respectively. They provide d.c. stability as well as a small signal negative feedback to improve the bandwidth of the amplifier.

D.C. amplifiers

In direct current or d.c. amplification, direct coupling is used as shown in Fig. 30.16. The base voltage of TR_2 is directly coupled to the collector of TR_1. The quiescent (no signal) condition of TR_2 is therefore determined by the previous stage. The absence of coupling capacitors makes d.c. amplifiers useful for low-frequency signals.

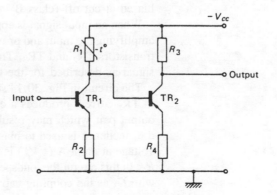

Fig. 30.16 Direct coupled (d.c.) amplifiers

D.C. amplifiers suffer from what is known as **drift**, which refers to a shift in the quiescent point of the amplifier caused by changes in temperature. To overcome the effect of drift, thermistors or other temperature sensitive elements may be used as shown in Fig. 30.16.

Feedback in amplifiers

Figure 30.17 shows a feedback system where a portion of the output voltage is fed back to the input. Voltage v_f is the fed-back voltage which is added to the overall input voltage v_i to produce the effective input voltage e_i. The feedback network B feeds all or a portion β of the output back to the input. If the output voltage is v_o then the fed-back voltage is

$$v_f = \beta v_o$$

The overall input to the amplifier is $v_i = e_i + v_f = e_i + \beta v_o$. Hence the gain with feedback is

$$G_f = \frac{v_o}{v_i} = \frac{v_o}{e_i + \beta v_o}$$

For negative feedback, where the fed-back voltage is in anti-phase to the input, the effective input $e_i = v_i - v_f$ giving a low overall gain.

For positive feedback where the fed-back voltage is in phase with the input, the effective input $e_i = v_i + v_f$ giving a high overall gain.

Using the values given in Fig. 30.17 and assuming negative feedback the following calculations may be made:

Effective input to the amplifier $e_i = 10 - 2 = 8$ mV

Given that the gain of the amplifier without feedback is $G = 100$, then

Output voltage $v_o = 8 \times 100 = 800$ mV

Hence the overall gain (with feedback) is

$$G_f = \frac{v_o}{v_i} = \frac{800}{10} = 80$$

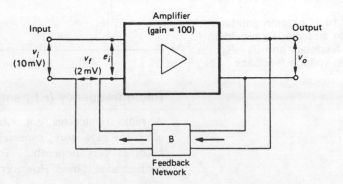

Fig. 30.17 Feedback in amplifiers

The percentage feedback β is

$$\beta = \frac{v_f}{v_o} = \frac{2}{800} = \frac{1}{400} = 0.0025 \text{ or } 0.25\%$$

Table 30.1 Comparison between positive and negative feedback

Positive feedback	Negative feedback
(1) High gain	(1) Low gain
(2) Narrow bandwidth	(2) Wide bandwidth
(3) Peaking frequency response	(3) Flat frequency response
(4) Low input impedance	(4) High input impedance
(5) High output impedance	(5) Low output impedance
(6) Introduces instability for both a.c. (hence possible oscillation) and for d.c. (hence unstable quiescent conditions)	(6) Improves stability for both a.c. and d.c.
(7) Used in oscillators	(7) Used frequently to improve stability and widen bandwith

Feedback may also be divided into current and voltage feedback. *Current feedback* is where the fed-back voltage is proportional to the output current, e.g. that due to R_4 in Fig. 30.18. Where the fed-back voltage is proportional to the output voltage, the feedback is known as *voltage feedback*, e.g. that due to $C_2 - R_3$ in Fig. 30.18.

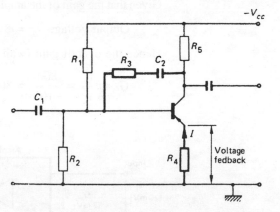

Fig. 30.18 Common emitter transistor amplifier R_4 providing current feedback and $C_2 - R_3$ providing voltage feedback

Radio frequency (r.f.) amplifiers

At radio frequencies, e.g. v.h.f. and u.h.f., the effect of the inter-electrode capacitors, especially that between collector and base, becomes very noticeable. To overcome this, the common base configuration is used. However, this configuration presents a low input impedance which loads the preceding stage feeding the amplifier. To rectify this, two methods may be used. The first is to use a common emitter amplifier with a neutralising circuit. Such circuits cancel out or neutralise the negative feedback effect of the c—b capacitance by introducing another feedback loop in opposition to it.

The second method is to use a common emitter amplifier in cascade with a common base amplifier (Fig. 30.19). TR_1 is the common emitter amplifier and TR_2 the common base amplifier. The input is fed to the base of TR_1 with its emitter decoupled to chassis by C_3. The output of TR_1 collector is fed to the emitter of TR_2 which has its base decoupled to chassis by C_1. The bias for TR_1 and TR_2 is provided by resistor chain $R_1 - R_2 - R_3$.

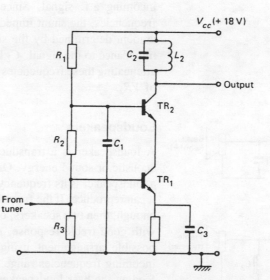

Fig. 30.19 Cascade amplifiers

Hi-fi

Hi-fi stands for high fidelity, a term applied to sound-producing equipment that gives a realistic re-creation of the original sound — in other words, high-quality reproduction. Hi-fi systems should therefore have a wide a.f. bandwidth 40 Hz—16 kHz, contain low noise and low distortion.

Tone control

The function of a tone control circuit is to extend or reduce (i.e. reshape) the frequency response curve of the amplifier with a bass control affecting the low-frequency end and a treble control affecting the high-frequency end. Tone control circuits may vary from a simple capacitor in series with a resistor to a very complex arrangement employing feedback. Fig. 30.20 shows a useful tone control circuit with a bass control for the low-frequency end and treble control for

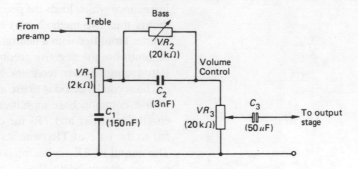

Fig. 30.20 Tone control circuit

the high-frequency end. $VR_1 - C_1$ offers a shunt impedance to the incoming a.f. signal. Since the reactance of C_1 is low at high frequencies, the shunt impedance attenuates high frequencies by an amount determined by the setting of VR_1. $VR_2 - C_2$ offers a series impedance to the signal. C_2 has a high reactance at low frequencies attenuating these frequencies by an amount determined by the setting of VR_2.

Loudspeakers

A loudspeaker is a transducer that converts electrical energy into acoustic or sound energy. One of the factors affecting the choice of a loudspeaker is its frequency response, i.e. the range of frequencies it can reproduce. If the frequency range of the loudspeaker is not wide enough, then two speakers, one with good bass response and another with good treble response, may be used. Figure 30.21 shows one possible arrangement using a cross-over network to divide the incoming frequencies range. It consists of a low pass filter $L_1 - C_1$ feeding the bass loudspeaker (known as a woofer) and a high pass filter $L_2 - C_2$ feeding the treble speaker (known as a tweeter).

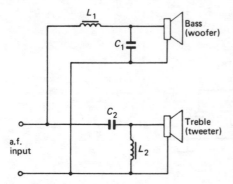

Fig. 30.21 Cross-over network feeding bass and treble loudspeakers

Other factors affecting the choice of the loudspeaker are output power, efficiency, and impedance for matching purposes.

31 The operational amplifier

The amplifiers considered so far are built up from separate discrete components, such as transistors, diodes and resistors. With an integrated circuit, all the necessary discrete components may be included in a single IC package. One such amplifier is known as the operational amplifier. Originally designed to perform certain mathematical operations, hence the name, the operational amplifier may be employed for a variety of applications.

The operational amplifier is an ideal amplifier with infinite gain, infinite bandwidth and a perfectly flat frequency response, infinite input impedance, zero output impedance, and no drift. In practice, an operational amplifier has the following properties:

Very high gain (over 50 000)
Very wide bandwidth and a flat frequency response
Very high input impedance
Very low output impedance
Very low drift

Figure 31.1 shows the symbol for an operational amplifier. It has two input terminals: an inverting input ($-$) which produces an output in anti-phase to the input, and a non-inverting input ($+$) which produces an output in phase with the input.

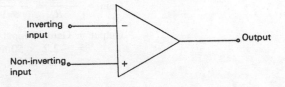

Fig. 31.1

Applications

There are a number of applications for an operational amplifier including an inverter, non-inverter, voltage follower, summing amplifier, integrator, differential amplifier and comparator. External

197

components are connected to the operational amplifier chip to determine its specific application. We will now consider some of these applications.

The inverting amplifier

Figure 31.2 shows an operational amplifier connected as an inverter.

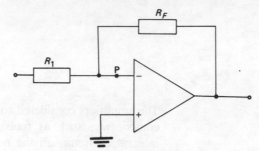

Fig. 31.2

With an almost infinite gain, the operational amplifier requires very small input to produce an output. Point P, the inverting input terminal is, therefore, at virtual earth, i.e. almost zero potential. To reduce the gain of the amplifier to the required level, a very large negative feedback is introduced by feedback resistor R_F. It can be shown that the gain of the inverter in Fig. 31.2 is

$$\text{Gain} = -\frac{R_F}{R_1}$$

The negative sign signifies the inverting operation.

EXAMPLE

For an input of 50 mV and given that $R_1 = 1 \text{ k}\Omega$ and $R_F = 2.2 \text{ k}\Omega$, calculate the gain and the expected output voltage.

Solution

$$\text{Gain} = -\frac{R_F}{R_1} = -\frac{2.2 \text{ k}\Omega}{1 \text{ k}\Omega} = -2.2$$

$$\begin{aligned}
\text{Output} &= \text{Gain} \times \text{Input} \\
&= -2.2 \times 50 \text{ mV} \\
&= -110 \text{ mV}
\end{aligned}$$

The summing amplifier

The summing amplifier (Fig. 31.3) provides an output voltage which is proportional to the sum of the input voltages V_1 and V_2. For input V_1, the gain is $-R_F/R_1$. For input V_2, the gain $= -R_F/R_2$.

For example, if $R_F = R_1 = R_2 = 5 \text{ k}\Omega$, then the gain for both inputs $= -5 \text{ k}\Omega/5 \text{ k}\Omega = -1$. Given $V_1 = +1 \text{ V}$ and $V_2 = +2 \text{ V}$ we get an output due to V_1 of $1 \times (-1) = -1 \text{ V}$ and an output due

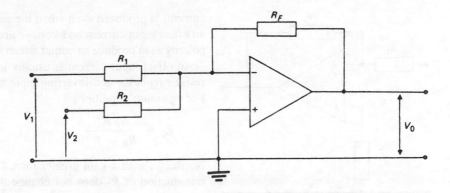

Fig. 31.3

to V_2 of $2 \times (-1) = -2$ V. Hence the total output V_o is $-1-2$ $= -3$ V.

EXAMPLE 1

In the operational amplifier shown in Fig. 31.4 input $V_1 = 20$ mV and input $V_2 = -10$ mV. Calculate V_o.

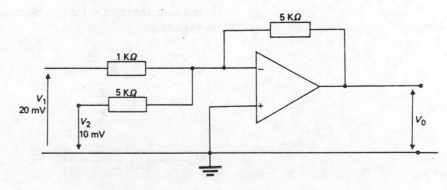

Fig. 31.4

Solution

The output due to V_1 is $-\dfrac{5}{1} \times 20 = -100$ mV.

The output due to V_2 is $-\dfrac{5}{5} \times (-10) = +10$ mV

Hence the total output $V_o = -100 + 10 = -90$ mV.

The voltage follower

In this case, the operational amplifier has a 100% negative feedback (Fig. 31.5) providing a gain of one. Note that the input and output are in phase.

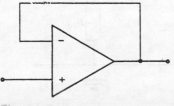

Fig. 31.5

Offset

For a zero input, an ideal operational amplifier has a zero output. In practice, this is not the case and a non-zero output (voltage and

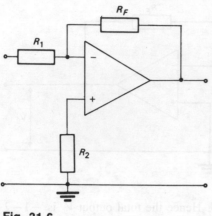

Fig. 31.6

current) is produced even when the input is zero. To overcome this, an offset input current and voltage are introduced of such value and polarity as to produce an output which will cancel the non-zero output.

An offset input current is usually introduced by the addition of a resistor R_2 in the non-inverting input terminal as shown in Fig. 31.6. For optimum conditions

$$R_2 = \frac{R_F R_1}{R_F + R_1}$$

Normally, with a gain greater than 4, R_2 is made equal to R_1. The introduction of R_2 does not change the gain of the inverter,

$$\text{namely } \frac{-R_F}{R_1}$$

In an IC package, Offset Null Connections are provided, as will be seen later.

The non-inverting amplifier

In this case, the input is fed into the non-inverting terminal as shown in Fig. 31.7.

$$\text{Gain} = \frac{R_1 + R_F}{R_1}$$

$$= 1 + \frac{R_F}{R_1}$$

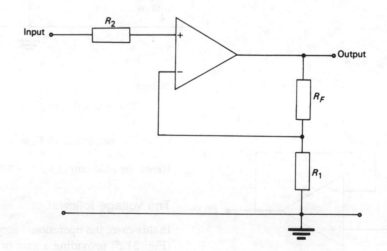

Fig. 31.7

The integrating amplifier

Figure 31.8 shows an operational amplifier connected as an integrator. The feedback component is, in this case, capacitor C_1. With a step input, the output is a ramp in the opposite polarity to the input, as

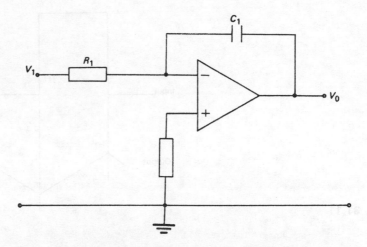

Fig. 31.8

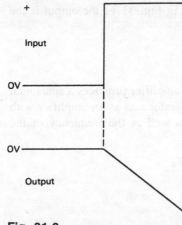

Fig. 31.9

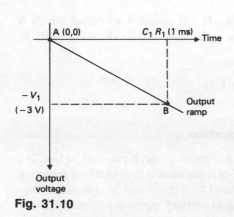

Fig. 31.10

shown in Fig. 31.9. The rate of change of the output is determined by the time constant $C_1 R_1$ as well as the level of the step input.

$$\text{Rate of change of output} = -\frac{V_1}{C_1 R_1} \text{ volts/second}$$

where V_1 is the input voltage. The negative sign denotes the inverting function.

EXAMPLE 2

(a) Given $C_1 = 1 \ \mu F$, $R_1 = 1 \ k\Omega$ and $V_1 = 3 \ V$, calculate the rate of change of the output.

Solution

Before calculating the rate of change of the output ramp, it is useful to find the time constant of the circuit.

$$\text{Time constant} = C_1 R_1 = 1 \times 10^{-6} \times 1 \times 10^{3}$$
$$= 10^{-3} \ s = 1 \ ms.$$

Rate of change of output

$$= -\frac{V_1}{C_1 R_1} = -\frac{3 \ V}{1 \ ms} = -3 \ V/ms$$

This means that the output will change by -3 V every 1 ms. Assuming that C_1 is fully discharged before the input is applied, then after 0.5 ms, the output is -1.5 V; after 1 ms, it is -3 V; after 1.5 ms, it is -4.5 V and so on. It is useful to note that the time constant of the integrating amplifier $C_1 R_1$ is the time it takes the output starting from zero to reach the same value as the input.

To draw a graph of the output ramp, two points are needed. The most appropriate points are A and B in Fig. 31.10. Point A where the output $V_o = 0$ at time $t = 0$. Point B where output voltage $V_o = -$input voltage and time $t = $ time constant $C_1 R_1$. Joining the two points A and B with a straight line gives the graph of the ramp output. In our previous example, A is 0, 0 and B is -3 V, 1 ms.

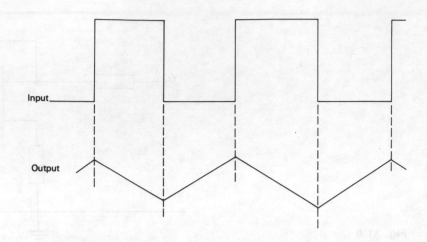

Fig. 31.11

For the square wave input shown in Fig. 31.11 the output is the waveform shown.

A.C. gain

A sinusoidal input to the integrating amplifier produces a sinusoidal output as well. In this case, the integrator acts as an amplifier with an a.c. gain determined by C_1R_1 as well as the frequency of the input.

$$\text{Gain} = \frac{\text{Capacitor reactance } X_{C_1}}{R_1}$$

$$\text{But } X_{C_1} = \frac{1}{2\pi f\, C_1}$$

$$\therefore \text{ Gain} = \frac{1}{2\pi f\, C_1 R_1}$$

The difference amplifier

The difference amplifier (Fig. 31.12) provides an output which is proportional to the difference in the input signals $(V_1 - V_2)$. With $R_1 = R_2$,

$$\text{Output } V_o = \frac{R_F}{R_1}\,(V_1 - V_2).$$

The 741 op amp pin allocation

Figure 31.13 shows the pin allocation of the 8-pin dual-in-line 741 op amp IC package. Zero offset adjustment is provided by connecting a variable resistor between Offset Null pins 1 and 5 (as shown in Fig. 31.14). The inverting and non-inverting inputs are at pins 2 and 3

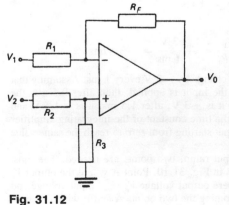

Fig. 31.12

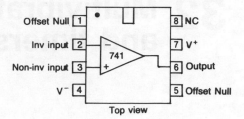

Fig. 31.13

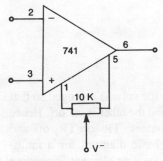

Fig. 31.14

respectively, while the output is taken at pin 6. A positive voltage (+15 V) is applied to pin 7 and an equal but negative voltage (−15 V) is applied to pin 4. Pin 8 is not used (No Connection).

Frequency compensation

Ideally an op amp has an infinite bandwidth. In practice, however, the gain drops as the frequency increases. To compensate for this fall, and thus increase the bandwidth, some IC packages provide frequency compensation. In the 748 op amp package, pins 1 and 8 are used for this purpose (Fig. 31.15). Compensation is achieved by connecting a capacitor C_1 as shown in Fig. 31.15(b).

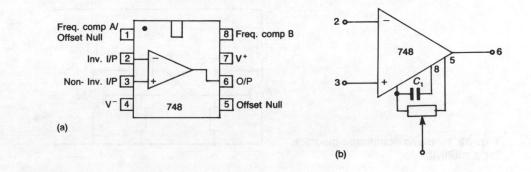

Fig. 31.15

32 Multivibrators and timers

Multivibrators consist of two transistors (or valves) arranged so that one transistor is always fully conducting and the other is cut off. Hence a multivibrator has two distinct stable states: TR_1-on/TR_2-off and TR_1-off/TR_2-on. Figure 32.1 shows a basic diagram for a multivibrator circuit where z_1 and z_2 are two coupling elements providing positive feedback.

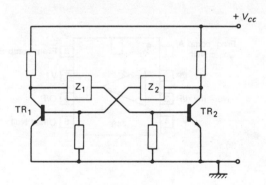

Fig. 32.1 Basic schematic diagram for a multivibrator

Depending on the feedback element used, multivibrators may be divided into three types.

1. The **bistable** which can stay permanently stable in either state. An external pulse may be used to switch it over from one state to the other. It then remains stable in that second state until triggered by a second pulse and so on.
2. The **monostable** which has one permanent state only. If triggered by an external pulse, it switches over to the other state. It remains in that second state for a duration determined by the time constant of the feedback elements, and returns back to the original state by itself.
3. The **astable** which is a free running oscillator having no permanent state. It continually changes from one state to the other and back again and so on.

The bistable circuit

Referring to Fig. 32.2, when the d.c. supply is switched on, because of component tolerances, etc. one transistor will tend to conduct more than the other. However small the difference in transistor currents, it is enough to define the state in which the bistable will settle in.

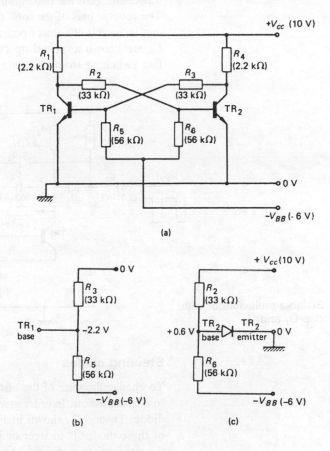

Fig. 32.2 Bistable multivibrator (a) Circuit using separate d.c. supplies V_{CC} and $-V_{BB}$

Supposing that TR_2 begins to conduct more than TR_1. TR_2 collector voltage will drop causing a drop in TR_1 base voltage, TR_1 current decreases, its collector voltage increases causing TR_2 base to go up away from the emitter. TR_2 current increases further and so on until TR_2 is fully on (saturated) and TR_1 is fully off. In this state TR_1 being at cut off has a collector voltage of $+V_{cc}$ (10 V), while TR_2 being at saturation has a collector voltage of zero volts. The voltage at the base of TR_1 is determined by resistor chain $R_3 - R_5$. As shown in Fig. 32.2(b), the base is kept at a negative potential by negative supply $-V_{BB}$ keeping TR_1 off. The voltage at the base of TR_2 is determined by resistor chain $R_2 - R_6$. As shown in Fig. 32.2(c) TR_2 base has a positive voltage, giving TR_2 the necessary forward bias. The bistable will thus remain in this state permanently if undisturbed by an external pulse. It may remain equally as stable in the other state if TR_1 started conducting more than TR_2 when first switched on.

To avoid using a separate negative d.c. supply, the circuit in Fig. 32.3 is used with R_6 as a common emitter resistor. In either state a voltage V_e develops across R_6 due to the current taken by the 'on' transistor. Thus the two emitters are kept at a constant voltage of V_e. The reverse bias of the 'off' transistor is now achieved by ensuring that its base is at a lower potential than its emitter. Capacitors C_2 and C_3 are known as **speed-up capacitors**. Their purpose is to ensure fast switching from one state to the other.

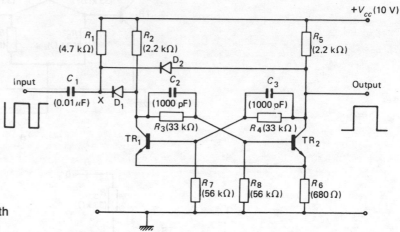

Fig. 32.3 Bistable multivibrator with steering diodes D_1 and D_2

Steering diodes

To change the state of the bistable, a pulse is fed into it turning the 'on' transistor off. In order to avoid using two separate inputs, steering diodes D_1 and D_2 shown in Fig. 32.3 are employed. The function of these diodes is to steer or guide the trigger pulse to the base of the appropriate transistor. Assuming that the bistable is in the state TR_1-off/TR_2-on, then point X, the common cathodes of D_1 and D_2, has a potential of V_{cc} (10 V). The anode of D_1 is also at V_{cc} (10 V) giving the diode a bias voltage of zero. On the other hand, D_2 anode is at emitter potential (TR_2 being at saturation) of approximately 1 V giving D_2 a reverse bias of -9 V.

If a negative pulse is now applied at point X, D_1 will conduct directing the pulse through R_3 to the base of TR_2 turning it off. As TR_2 turns off, TR_1 switches on and the bistable settles in its alternative state. The bias of the steering diodes is now opposite to what it was with D_1 having a reverse bias of -9 V. A second pulse will forward bias D_2 directing the pulse through R_4 to the base of TR_1 turning it off and switching the bistable back to its original state.

As the bistable continually switches over, a square wave output may be taken at the collector of either transistor.

Input

Waveform
at point x

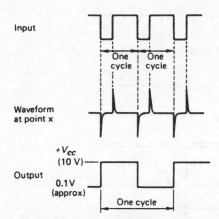

$+V_{cc}$
(10 V) ——

Output

0.1 V
(approx)

One cycle

Fig. 32.4 Input and output
waveforms of a bistable multivibrator

Figure 32.4 shows the input and output waveforms of the bistable
described above. The square wave input is first differentiated by
$C_1 - R_1$. The negative spikes thus produced are used to trigger the
bistable, producing the output shown. The positive-going spikes have
no effect on the circuit as both diodes will be reverse biased. The
output is another square wave having a frequency half that of the input.
This is why the bistable is known as a divide by two ($\div 2$) device
which is extensively used in counters and calculators.

The monostable

The feedback loop of the monostable (Fig. 32.5) contains one $C - R$
network, $C_2 - R_2$. When the circuit is first switched on TR_2, having
its base taken towards V_{cc} via R_2, conducts very heavily, turning TR_1
off in the process. The negative d.c. supply $-V_{BB}$ ensures that TR_1
remains off. This is the permanent state of the circuit.

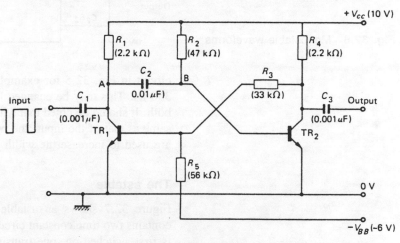

Fig. 32.5 Monostable multivibrator

If a negative pulse is now applied at the input, the high-frequency
edge will pass through C_2 to the base of TR_2 turning it off. In a
similar way to the bistable, TR_1 will saturate and TR_2 will cut off.
The collector of TR_1 (point A) previously at $+10$ V (V_{cc}) goes down
sharply to zero volts, a drop of 10 V giving capacitor C_2 a charge
of -10 V. In other words TR_2 base (point B) is now at -10 V
holding the transistor off. Capacitor C_2 begins to discharge through
R_2 from -10 V, attempting to charge up to $+10$ V. The negative
potential at point B begins to decrease gradually at a rate determined
by time constant $C_2 R_2$. As point B crosses the zero line (Fig.
32.6(b)), TR_2 begins to conduct turning TR_1 off, and the monostable
is back to its original state, awaiting the second trigger pulse.

The output of the circuit is a pulse waveform shown in Fig. 32.6(c).
The pulse width is determined by the period during which TR_2
remains in the off state, itself determined by time constant $C_2 R_2$. The

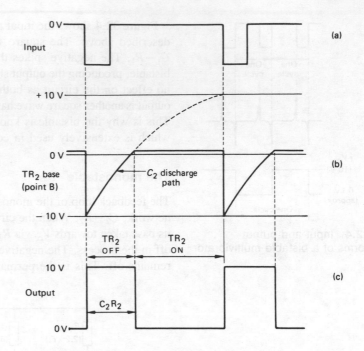

0 V

Input

+ 10 V

0 V

TR₂ base
(point B)

– 10 V

C₂ discharge path

TR₂ OFF

TR₂ ON

10 V

Output

C₂R₂

0 V

(a)

(b)

(c)

Fig. 32.6 Monostable waveforms

circuit in Fig. 32.5 for example gives a pulse duration of approx. 350 μs. This may be changed by varying the value of C_2 or R_2 or both. It should be noted that while the frequency of the output is the same as that of the input, the pulse width is different. Monostables are used to increase the width of a pulse and for time delay.

The astable

Figure 32.7 shows an astable multivibrator. The feedback loop contains two time constant circuits C_1R_1 and C_2R_2. When the circuit is first switched on, one transistor tends to conduct more than the other. Due to the feedback loop this results in one transistor fully on while the other is fully off. Suppose that TR_1 turns on and TR_2 off with C_1 charging up to $-V_{cc}$ keeping TR_2 off. Capacitor C_1

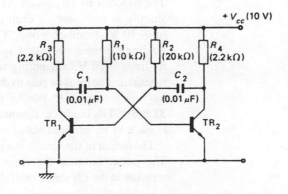

+ V_{cc} (10 V)

R_3
(2.2 kΩ)

R_1
(10 kΩ)

R_2
(20 kΩ)

R_4
(2.2 kΩ)

C_1
(0.01 μF)

C_2
(0.01 μF)

TR₁

TR₂

Fig. 32.7 Astable multivibrator

begins to discharge through R_1 attempting to charge up to $+V_{cc}$ as in the case of the monostable. As the potential of the junction $R_1 - C_1$ (TR$_2$ base) passes through zero, TR$_2$ switches on, turning TR$_1$ off. C_2 now charges up in the negative direction keeping TR$_1$ off. As C_2 discharges through R_2 it turns TR$_1$ on and so on. The circuit oscillates and produces a square wave output at either collector. The mark-and-space times are determined by the time constants of the circuit.

Figure 32.8 shows the output from the collector of each transistor. For equal mark-and-space, C_1R_1 and C_2R_2 are made equal. Note that as in the case of the monostable, the *CR* time constant feeding into the base of a transistor determines the time that the transistor remains off.

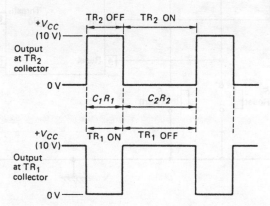

Fig. 32.8 Outputs from the collector of each transistor of the astable in Fig. 32.7

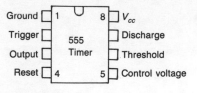

Fig. 32.9

The 555 timer

The 555 timer (Fig. 32.9) is a highly versatile low-cost IC that is specifically designed for precision timing applications, but which can also be used in a variety of monostable, astable and trigger applications.

When used in the timing mode, the 555 IC produces very accurate timing periods which may be varied from a few microseconds to several hundred seconds. The timing period is controlled by a single *CR* network and is virtually independent of the actual d.c. supply voltage. The timing period can be started via a TRIGGER pulse to pin 2 and aborted by a RESET command to pin 4.

A basic timer circuit is shown in Fig. 32.10 in which C_1R_1 is the timing network. The timing period is initiated by closing push button switch S_1 taking pin 2 to 0 V thus triggering the IC. Capacitor C_1 begins to charge up via resistor R_1 and the output at pin 3 goes up to a positive voltage. After a period of time of 1.1 × time constant C_1R_1, the output is reset to 0 V (Fig. 32.11). The timing period may be aborted by closing RESET press button switch S_2 causing pin 4 to go to 0 V. Under normal operation, S_2 is open and pin 4 is at V_{cc}.

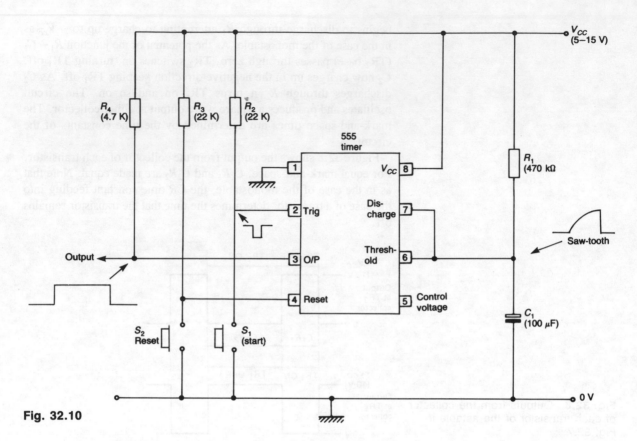

Fig. 32.10

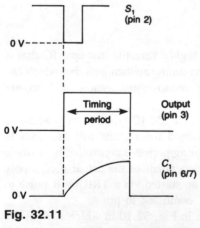

Fig. 32.11

33 Oscillators

The oscillator is an amplifier with adequate positive feedback to produce an output without an external input. It converts d.c. (the supply to the amplifier) to a.c. signal. Two basic requirements are necessary for sustained oscillations:

(a) positive feedback and
(b) a gain round the feedback loop of more than one.

There are two types of oscillators: the sinusoidal oscillator producing sine wave outputs; and the non-sinusoidal, also known as relaxation oscillator or multivibrator, normally producing square waves.

Tunes collector oscillators

In Fig. 33.1, $L_2 - C_2$ is the tuned circuit across which the output is developed. Part of the output is fed back to the input via transformer coupling $L_2 - L_1$ in such a way as to be in phase with the input. The transistor is connected in the common emitter mode with $R_1 - R_2$ providing class A biasing. C_1 is a bias decoupling capacitor for R_2. R_3 is the normal d.c. stabilising resistor with C_3 its decoupling capacitor.

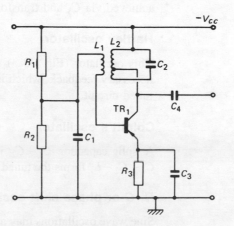

Fig. 33.1 Tuned collector oscillator

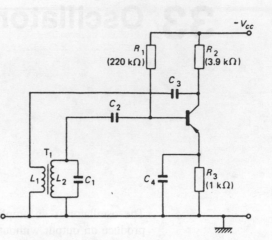

Fig. 33.2 Tuned base oscillator

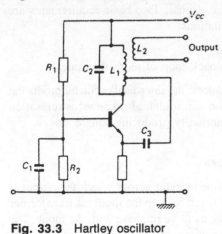

Fig. 33.3 Hartley oscillator

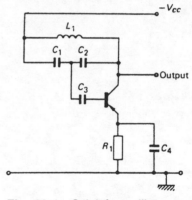

Fig. 33.4 Colpitt's oscillator

Tuned base oscillators

In Fig. 33.2, C_2 is a coupling capacitor which provides class C biasing for TR_1. L_2-C_1 is the tuned circuit. Positive feedback is achieved via C_3 and transformer T_1.

Hartley oscillator

This oscillator (Fig. 33.3) uses a split inductor L_1 to provide the necessary feedback, which is fed into the emitter. C_2-L_1 forms the tuned circuit.

Colpitt's oscillator

A split capacitor C_1-C_2 is used in this case as in Fig. 33.4. $C_1-C_2-L_1$ forms the tuned circuit, with C_3 providing class C bias.

R–C or phase shift oscillators

Sine wave oscillations may also be produced by using a suitable $R-C$ feedback network as shown in Fig. 33.5. R_1-C_1, R_2-C_2 and

$R_3 - C_3$ form a phase shift network which at one given frequency provides 180° phase shift. Since the transistor itself provides 180° phase shift, a total of 360° phase shift is produced round the feedback loop. The feedback is, therefore, positive. Normally $R_1 - C_1$, $R_2 - C_2$ and $R_3 - C_3$ are made equal with each giving a phase shift of 60°. It should be noted that the $R - C$ network provides 180° phase shift at one frequency only, determined by the value of the components.

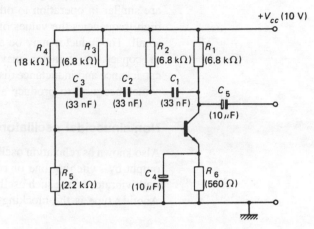

Fig. 33.5 Phase shift oscillator, $R_1 - C_1$, $R_2 - C_2$; $R_3 - C_3$ provide 180° phase shift

Crystal oscillators

One very important requirement of an oscillator is the stability of its frequency. Changes in frequency may be caused by changes in the values of the tuned circuit components L and C or changes in transistor parameters due to temperature variations. Frequency stability may

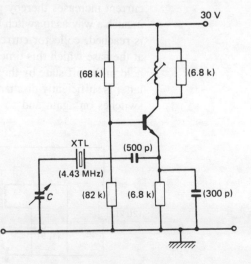

Fig. 33.6 Crystal oscillator. The circuit is used as the reference oscillator in TCE 3000 colour TV receiver. Note that the self-capacitance of the inductor is used for tuning. Variable capacitor C is in fact a varactgor (By permission of Thorn Consumer Electronics)

be improved by the proper choice of the circuit elements including the transistor. For very high stability a quartz crystal is used as shown in Fig. 33.6, with the crystal determining the frequency of oscillation.

However, small variations may be achieved by placing a variable capacitor C across the crystal. Crystal oscillators are used in colour TV receivers to produce the 4.43 MHz sub-carrier to an accuracy of a few Hz.

U.H.F. oscillators

Oscillators at very high (v.h.f.) and ultra high (u.h.f.) frequencies are similar in operation to other oscillators. However, due to such high frequencies the values of tuning components L and C are very small. The inductor may be a single strip of wire or a simple loop of copper. A varactor may be used for the capacitance. The capacitance and inductance displayed by a short transmission line are sometimes used to produce a tuned circuit.

Non-sinusoidal oscillators

Also known as relaxation oscillators, these produce a square or pulse output by switching one or two transistors on and off. The astable multivibrator is one such oscillator described in the previous section. Another type is the blocking oscillator.

Blocking oscillators

This type uses a transformer coupled feedback from collector to base (Fig. 33.7). The action of the circuit hinges around the fact that, due to transformer coupling, a voltage is induced into the base only when collector current is varying, i.e. increasing or decreasing. In the one case the feedback is positive and in the other it is negative. When the circuit is first switched on the transistor conducts and collector current increases thereby producing a feedback voltage at the base in such a way as to switch the transistor further on. When saturation is reached, collector current ceases to increase, inducing a voltage at the base which this time turns the transistor off. The transistor is held in the off state by the negative charge on capacitor C, until the latter is sufficiently discharged through resistor R when the transistor switches on again and so on.

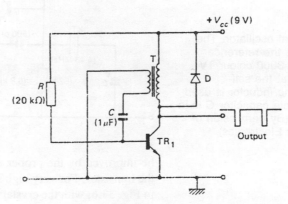

Fig. 33.7 Blocking oscillator

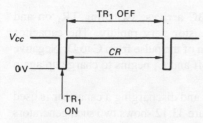

Fig. 33.8 Output of a blocking oscillator

The output from a blocking oscillator is a narrow pulse waveform as shown in Fig. 33.8. The width or mark of the pulse is determined by the transformer parameters while the space is determined by time constant CR. The frequency of oscillation may, therefore, be varied by changing the value of resistor R.

The secondary winding of the transformer acts as the collector load. The rapid change in current through the winding when the transistor switches off causes a large back e.m.f. to be built up resulting in a large overshoot in collector voltage as the transistor switches off. This overshoot voltage may be of such a magnitude as to exceed the maximum rateable collector voltage resulting in damage to the transistor. To protect the transistor a diode D_1 is connected across the primary winding of the transformer as shown. The diode is normally reverse biased. It is forward biased only if the collector voltage exceeds that of the d.c. supply V_{cc}. This way the voltage at the collector cannot exceed that of the supply.

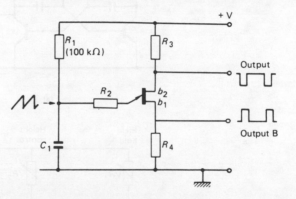

Fig. 33.9 Unijunction oscillator

Unijunction oscillators

Devices having negative resistance characteristics such as the unijunction readily lend themselves for use in oscillators. Figure 33.9 shows an oscillator using a unijunction. The unijunction is biased to operate along that part of its output characteristics where the output current increases as the output voltage decreases, i.e. negative resistance. It continually switches on and off without any feedback. The output at base 2 (b_2) is a pulse waveform. Another output may be obtained from base 1 (b_1) which is a positive-going pulse waveform. A saw-tooth output may also be taken at the emitter. The periodic time or frequency is determined by time constant C_1R_1.

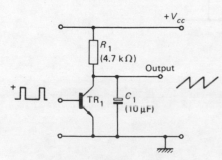

Fig. 33.10 Saw-tooth waveform produced from a pulse input

Saw-tooth generators

Figure 33.10 shows a saw-tooth generator using a pulse input. For the period of the input cycle between A and B (Fig. 33.11) the base of the transistor is at zero voltage and the transistor is cut off. Capacitor C_1 thus steadily charges up through resistor R_1. Before the capacitor

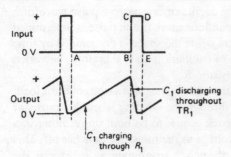

Fig. 33.11 Input and output waveforms of the saw-tooth generator in Fig. 33.10

is fully charged, positive edge BC arrives, switching TR_1 on and discharging C_1 through the transistor very rapidly. The capacitor remains discharged for the duration of the pulse from C to D. Negative edge DE switches the transistor off and C_1 begins to charge up again and so on.

The same principle of charging and discharging a capacitor is used in other saw-tooth generators. Figure 33.12 shows two such generators using an astable multivibrator and a blocking oscillator respectively. Such oscillators are used in TV receivers for the field timebase. VR_1 controls the frequency (field hold) and VR_2 controls the amplitude (height) of the waveform.

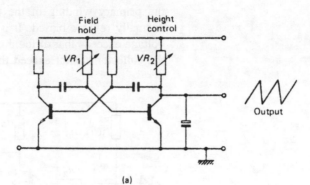

(a)

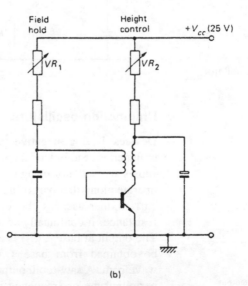

(b)

Fig. 33.12 Saw-tooth generators used in the field timebase of TV receivers. (a) Using an astable multivibrator (b) Using a blocking oscillator

34 Logic circuits

Logic elements, including gates, counters and memory devices, are manufactured in IC packages. These ICs are classified into categories, known as families, according to the number of gates or equivalent elements that they contain. These families are:

small scale integration (SSI): up to 10 gates
medium scale integration (MSI): 10—100 gates
large scale integration (LSI): 100—1000 gates
very large scale integration (VLSI): 1000—10 000 gates
super large scale integration (SLSI): 10 000—100 000 gates

The level of integration represents the complexity of the IC package. It increases in powers of 10, i.e. 10, 100, 1000 and so on. Small and medium scale integration (SSI and MSI) provide discrete logic elements such as gates, counters and registers. Large and very large scale integration (LSI and VLSI) provide memory chips, microprocessors and complete systems such as microcomputers.

Logic states

A logic element has two separate states: logic 0, represented by low voltage, normally 0 V; and logic 1, represented by a high voltage (a positive voltage in the case of **positive logic** and a negative voltage in the case of **negative logic**). The voltage level which represents logic 1 depends on the type used. For bipolar technology, e.g. TTL (transistor—transistor logic), logic 1 is represented by 5 V, while for CMOS (complementary metal oxide silicon) logic, 1 may vary between 3 V and over 15 V. While TTL has the advantage of speed, CMOS has a higher component density (i.e. more gates per IC package) and does not require a stabilised d.c. supply.

Transistor—transistor logic (TTL)

This type of logic is used in integrated circuits and has the advantage of fast switching action. Figure 34.1 shows a simplified NAND gate using a multiple emitter transistor TR_1 at the input. With both inputs at logic 0, TR_1 saturates, making its collector go down to

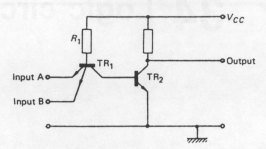

Fig. 34.1 NAND gate (TTL circuit)

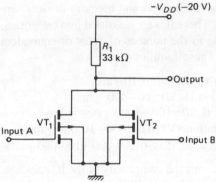

Fig. 34.2 NOR gate (fet logic)

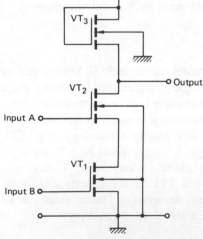

Fig. 34.3 NAND gate (fet logic)

approximately 0 V. Hence TR$_2$ is off and the output is 1. With both inputs at logic 1, TR$_1$ turns off, switching TR$_2$ on to give an output of 0.

FET logic gates

Logic gates are now manufactured as integrated circuits, ICs. A large number of gates may then be constructed on a tiny chip of silicon of 1 mm × 2 mm. Because of their simplicity, fets are more commonly used than bipolar transistors. One type of fet logic is known as CMOS. Figure 34.2 shows a NOR gate operated by negative logic. VT$_1$ and VT$_2$ are p-channel mosfets of the normally-off (enhancement) type. When both inputs are at logic 0 level, VT$_1$ and VT$_2$ are cut off giving an output of 1 ($-V_{DD} = -20$ V). When either or both inputs are at logic 1 level (say -20 V) one or both transistors are on, giving an output of 0.

Figure 34.3 shows a NAND gate using an n-channel normally-off mosfet. Positive d.c. supply is used and positive logic is necessary to operate the gate. VT$_3$ is biased permanently on by taking its gate to V_{DD} and acts as a load to the gate. When one or both inputs are at 0 level, one or both transistors are off giving a 1 output. Only when both inputs are 1 will current flow giving an output of 0.

Boolean expressions

The functions of the individual gates or combinations of gates may be expressed in a logical statement known as a Boolean expression. Boolean algebra uses the following notations (see Table 34.1):

- The AND function is represented by the dot symbol (.). Thus a 2-input (A, B) AND gate has an output represented by the Boolean expression A.B.
- The OR function is represented by the plus symbol (+). Thus a 2-input OR gate has an output represented by A+B.
- The NOT function is represented by a bar over the input. Thus a NOT gate with input A has an output represented by $\overline{A}$ (read as NOT A).

From the above, it follows that:

- The NAND function is expressed by $\overline{A.B}$ (NOT-AND).
- The NOR function is expressed by $\overline{A+B}$ (NOR-OR).
- The Ex-OR function is given as $(A.\overline{B}) + (\overline{A}.B)$. It is also given a special notation $A \oplus B$.
- The Ex-NOR function is given as $\overline{(\overline{A}.\overline{B}) + (A.B)}$. It is also represented as $\overline{A \oplus B}$.

Table 34.1 Boolean expressions

Function	Boolean notation
AND	A.B
OR	A + B
NOT	$\overline{A}$
NAND	$\overline{A.B}$
NOR	$\overline{A+B}$
Ex-OR	$(A.\overline{B}) + (\overline{A}.B)$ or $A \oplus B$
Ex-NOR	$(\overline{A}.\overline{B}) + (A.B)$ or $\overline{(A \oplus B)}$

Combinational logic

Consider the logic circuit shown in Fig. 34.4. The function of the circuit may be worked out using the Boolean expressions as follows:

Output of OR gate (i) is $A+B$
Output of NAND gate (ii) is $\overline{C.D}$
∴ Output of OR gate (iii) is $(A+B) + (\overline{C.D})$

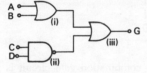

Fig. 34.4

EXAMPLE 1

Refer to Fig. 34.5

(a) Find the Boolean expression of the function of the combination of gates.
(b) Construct the truth table showing the logic states at all points and hence prove that the combination may be reduced to a single gate.

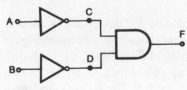

Fig. 34.5

Solution

(a) Boolean expression at C = $\overline{A}$
Boolean expression at D = $\overline{B}$
Boolean expression at F = $\overline{A}.\overline{B}$

(b) Truth table.

Inputs				Output
A	B	C	D	F
0	0	1	1	1
0	1	1	0	0
1	0	0	1	0
1	1	0	0	0

As can be seen from the truth table, the output function is the same as that of a NOR gate.

EXAMPLE 2

Refer to Fig. 34.6.

(a) Find the Boolean expression of the function of the combination of gates.
(b) Construct the truth table showing the logic states at all points and hence prove that the combination may be reduced to a single gate.

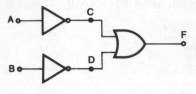

Fig. 34.6

Solution

(a) Boolean expression at C = $\overline{A}$
Boolean expression at D = $\overline{B}$
Boolean expression at F = $\overline{A} + \overline{B}$

(b) Truth table

Inputs				Output
A	B	C	D	F
0	0	1	1	1
0	1	1	0	1
1	0	0	1	1
1	1	0	0	0

From the truth table, it can be seen that the combination gate circuit is equivalent to a NAND gate.

Counters

The basic binary counter is a bistable or $\div 2$ device also known as a **flip-flop**. A number of these elements constitute a counter. Figure 34.7 shows a binary counter consisting of three flip-flops. Each flip-flop is capable of dividing incoming pulses by a factor of 2. Two successive flip-flops will thus divide by $2 \times 2 = 4$ ($\div 4$), and three flip-flops divide by $2 \times 2 \times 2 = 8$ ($\div 8$). In other words, for every 8 input pulses, 4 appear at output A, 2 at output B, and 1 at output C, as shown in Fig. 34.7.

As explained in Chapter 32, only one edge of the input pulse causes the bistable to change its state. The input changes from 0 to 1 and back to 0 and so on for every pulse. It is normally assumed that the negative-going edge of the pulse (i.e. from 1 to 0) changes the state of the flip-flop.

Output Q_A from bistable A represents the 2^0 or units column, output Q_B the 2^1 or twos column, and output Q_C the 2^2 or fours column (Table 34.2). Hence after 6 pulses the outputs are as follows: A (units) is 0, B (twos) is 1, and C (fours) is 1, making $0+2+4 = 6$. In binary coding this is written as 110 in the order CBA. It should

Table 34.2

Pulse	Q_A units (2^0)	Q_B twos (2^1)	Q_C fours (2^2)
0	0	0	0
1	1	0	0
2	0	1	0
3	1	1	0
4	0	0	1
5	1	0	1
6	0	1	1
7	1	1	1
8	0	0	0

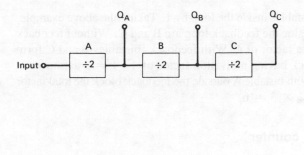

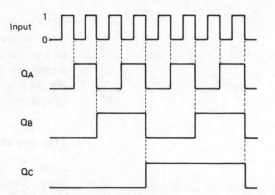

Fig. 34.7 3-element binary counter

be noted that the units column is the first bistable starting from the input. In binary coding on the other hand the units column is always on the right-hand side.

At the seventh pulse all outputs are 1. Pulse number 8 resets all flip-flops back to 0.

Note that the output of each flip-flop represents a column in the binary code. The binary code itself is in the order of CBA.

Table 34.3

Pulse	Q_A	Q_B	Q_C
0	0	0	0
1	1	0	0
2	0	1	0
3	1	1	0
Feedback	(0)	(0)	(1)
4	0	1	1
5	1	1	1
6	0	0	0

Feedback

Feedback may be applied to binary counters in order to change the factor by which the input is divided. For example, consider the feedback applied to the 3-element counter shown in Fig. 34.8(a). The count proceeds normally up to the third pulse when outputs CBA are 011 (see Table 34.3). The fourth pulse changes A to 0, B to 0, and C to 1. Without feedback, output CBA would be 100. However with feedback the change in C is fed back to B, changing it back to 1, giving CBA reading of 110. Pulse 5 then produces a reading of 111 and the 6th resets all elements back to 0. Thus we have a ÷6 counter.

In general it can be shown that a feedback loop reduces the division

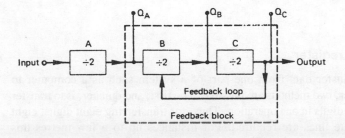

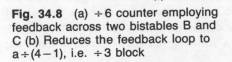

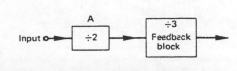

(a)   (b)

Fig. 34.8 (a) ÷6 counter employing feedback across two bistables B and C (b) Reduces the feedback loop to a ÷(4−1), i.e. ÷3 block

factor of the bistables inside the loop by 1. Taking the above example, the bistables inside the feedback loop are B and C. Without feedback they divide by a factor of 4. With feedback, bistables B and C form a feedback block having a division factor of $4 - 1 = 3$ as shown in Fig. 34.8(b). With bistable A outside the feedback block the total factor of division is $2 \times 3 = 6$.

The decimal counter

Figure 34.9(a) shows a decimal counter employing two feedback loops. Two feedback blocks are thus formed: block 1 enclosing bistables C and D, dividing by $4 - 1 = 3$; block 2 encloses block 1 together with bistable B. Referring to Fig. 34.9(b), without feedback block 2 would divide by $2 \times 3 = 6$. With feedback the division factor is $6 - 1 = 5$. With bistable A outside the feedback loops, the total division factor is thus $2 \times 5 = 10$.

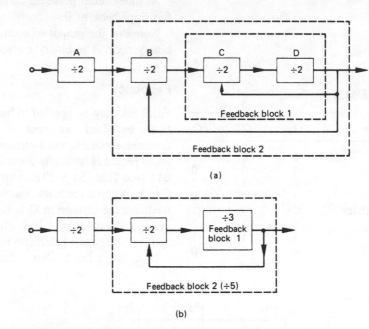

Fig. 34.9 (a) Decimal counter
(b) Reduction to feedback blocks

Shift register

To transfer data from one part of a system such as a computer to another, two methods are possible. The first, and quicker, is to transfer all the digits in one operation. Thus, for transferring eight digits, eight separate lines are required. For distances up to a few metres this method is used, but for longer distances such as data transmission between cities this method becomes too expensive. A second but slower method is used. The data information are transmitted digit by digit along a single wire. A shift register may be used to move the binary digits one stage to the left or to the right. The shift register consists of a number of flip-flops capable of rearranging the binary digits in successive order.

Ring counter

A ring counter consists of a standard counter comprising a number of bistables in which the output is taken back to the input — hence its name. Pulses may thus circulate round the counter from the input to the output and back to the input and so on. At the end of each cycle the output may be taken to trigger another counter. For example the output of a scale-of-ten or decade counter may be used to trigger another decade counter to make a scale of 100.

Latch

Bistables also known as flip-flops were considered in Chapter 32 and a circuit using discrete components was identified. Bistables are very important and useful devices in logic application. They are used as divide by two units, a **latch** (a single bit memory) or for other applications.

A basic bistable known as the $S-R$ bistable or latch is shown in Fig. 34.10. Two outputs are provided Q and $\bar{Q}$ (NOT Q) which are in anti-phase to each other. When $Q = 1$, then $\bar{Q} = 0$ and vice versa.

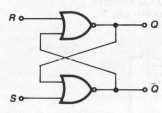

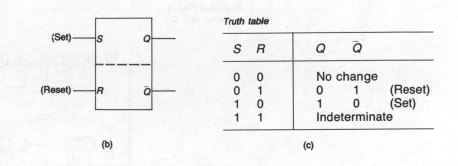

Fig. 34.10

(a)   (b)   (c)

The basic truth table is that shown in Fig. 34.10. A logic 1 on the reset R input forces Q to 0 (and $\bar{Q}$ to 1) while a logic 1 at the set S input forces Q to 1 (and $\bar{Q}$ to 0).

A development of the simple $S-R$ bistable is the clocked $J-K$ bistable which includes a clock line, and has no indeterminate state (Fig. 34.11). The clock line is essential for synchronous systems, in which case any change-over occurs only when the clock line is at logic 1. With both J and K at logic 1, the output toggles between logic 0 and logic 1 at every clock pulse.

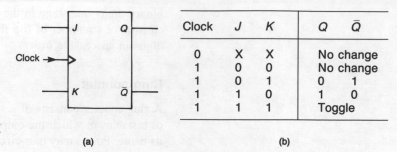

Clock	J	K	Q	Q̄
0	X	X	No change	
1	0	0	No change	
1	0	1	0	1
1	1	0	1	0
1	1	1	Toggle	

Fig. 34.11 (a) J–K symbol (b) J–K truth table

(a)

(b)

IC packaged gates

Gates are manufactured in integrated circuit form and are provided in packages of a number of identical gates on a single chip. Two main types used are: TTL and CMOS. Examples of TTL packages are given in Fig. 34.12.

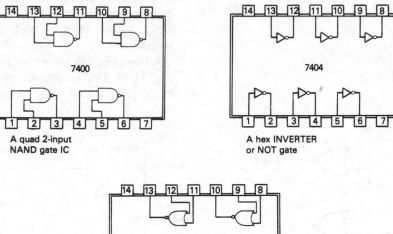

A quad 2-input
NAND gate IC

A hex INVERTER
or NOT gate

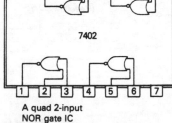

A quad 2-input
NOR gate IC

Fig. 34.12

35 Digital display

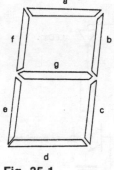

Fig. 35.1

There are several methods by which the readings of a digital instrument may be displayed ranging from the neon tube to the light emitting diode (LED) and the liquid crystal display (LCD). The number of digits displayed is usually between four and eight. A four-digit display gives a maximum indication of 9999, while an eight-digit display gives a maximum indication of 99999999. For cheaper digital meters, a display known as three-and-a-half- digit display is used. In this type, the most significant digit, MSD (the digit on the extreme left), can only display 0 or 1. Thus the maximum indication is 1999.

The two main types of displays are the light emitting diode (LED) and the liquid crystal display (LCD). Both consist of seven segments, being the minimum number of segments to represent the decimal numbers from zero to nine. The segments are labelled a to g, as shown in Fig. 35.1. Figure 35.2 shows how this type of display may be used to display decimal numbers. The liquid crystal displays require very little power (in μW) to operate, and are used in portable meters. LED

Fig. 35.2

Input count	Segments activated	Resulting display 0
0	a b c d e f	0
1	b c	1
2	a b d e g	2
3	a b c d g	3
4	b c f g	4
5	a c d f g	5
6	c d e f g	6
7	d e f g	7
8	a b c d e f g	8
9	a b c f g	9

displays require more current than LCD and consume more power (in mW). They are widely used in bench DVMs, where power consumption is not a primary consideration. On the other hand, LED displays can be observed in the dark while the liquid crystal display requires ambient light.

LED drive requirement

Figure 35.3 shows a drive system for one 7-segment LED display.

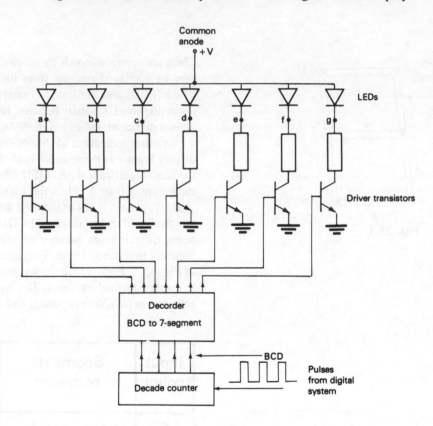

Fig. 35.3

The LED converts electrical current to light. Thus, to illuminate one segment of the display, current must be directed to the diode making up that segment. As shown, one terminal of each diode is connected to a common point. In the example shown, the anodes are connected together making what is known as common-anode configuration. If the cathodes were connected together, the result is a common-cathode configuration.

Each 7-segment display forms one digit of a complete display. Each digit thus has eight terminals, one for each segment and one common connection. In some instances, a decimal point is included in the display, giving a ninth terminal.

Before the segments are driven, the output of the decade (decimal) counter must be changed into the appropriate signal to drive the display. The output from the counter is in the form of binary coded decimal (BCD) which is a 4-line binary information. It must be converted into an appropriate signal to drive each of the 7 segments. This conversion is carried out by the decoder.

In practice, the switching driver transistors are contained in a single IC package. Furthermore, the functions of the decoder and the driver may be performed by a single IC, e.g. the 7447 for the common-anode and the 7448 for the common-cathode displays. Figure 35.4 shows a 2-digit display unit comprising two decade counters, two decoder/driver packages and two common anode 7-segment displays.

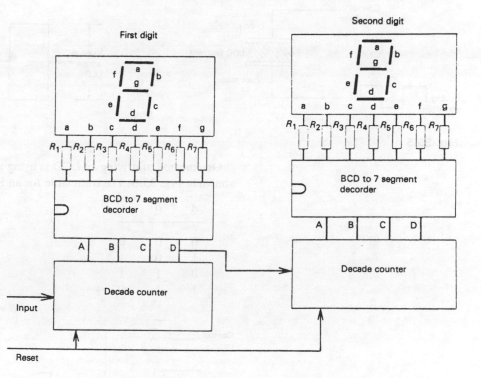

Fig. 35.4

LCD drive requirements

Liquid crystal displays utilise the unique interaction between electrical and optical characteristics of a family of liquids which retain their crystalline form. They give rise to optical properties which make them important for display operations. Since no light is generated (as is the case with LEDs), the display requires very small currents and consumes little power. Because of this, LCDs may be driven directly by MOS and CMOS circuitry.

LCD devices have two terminals or planes; the front plane FP and the back plane BP. The back planes of all segments are connected in common. Alternating voltages are always used for operating liquid crystal displays in order to avoid electrolytic plating and to ensure maximum life.

The method of driving a liquid crystal display involves feeding a low-frequency (typically 40 Hz) square wave to the back plane (Fig. 35.5). To turn a segment off an in-phase square wave is fed to the front plane and to turn a segment ON an out-of-phase square wave is fed to the front plane. When the driving and control waveforms are out of phase, the resultant waveform across the LCD is twice the amplitude of the driving square wave as shown in Fig. 35.5(b) which turns the segment ON.

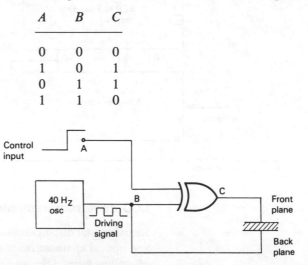

Fig. 35.5

One method of driving an LCD is using an Exclusive-OR gate, as shown in Fig. 35.6. The truth table for an Ex-OR gate is as follows:

A	B	C
0	0	0
1	0	1
0	1	1
1	1	0

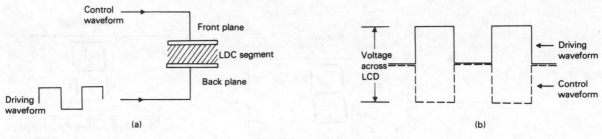

Fig. 35.6

From the truth table, it can be seen that when the control input A is 0, the output C follows the input in B, i.e. it is in phase with the 40 Hz driving signal at B. This in-phase signal is fed to the front plane of the LCD segment. At the same time, the back plane is directly fed with the 40 Hz driving signal, resulting in zero electric field across the segment. The segment will be OFF. When control input A is logic

1, the output C is out-of-phase with input B. This out-of-phase driving signal is fed into the front plane of the LCD segment. With the driving signal fed into the back plane, the resultant electric field across the segment is twice the 40 Hz driving signal. The segment will be ON.

36 Microprocessor applications

Advances made in integrated circuits and the introduction of large scale integration (LSI) have greatly increased the use of microprocessors in computers, control systems, instruments and data logging.

The basic elements of a microprocessor-based system (considered in Chapter 12) are reproduced in Fig. 36.1. The microprocessor, also known as the central processing unit (CPU) carries out the arithmetic and logic operations in accordance with a program stored in memory. The whole system is synchronised and controlled by the CPU.

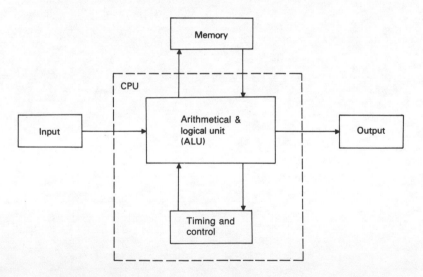

Fig. 36.1

Memory

There are two main types of memory chips. The Read Only Memory, ROM, is a permanent data store. The data is entered into ROM at the manufacturing stage, and remains there permanently. The CPU can only READ from ROM. It cannot WRITE into it, i.e. it cannot

Binary address				Binary data
0000	0000	0000	0000	DATA
0000	0000	0000	0001	DATA
0000	0000	0000	0010	DATA
0000	0000	0000	0011	DATA
0000	0000	0000	0100	DATA
1100	11J0	0110	1100	DATA
1100	1110	0110	1101	DATA
1111	1111	1111	1100	DATA
1111	1111	1111	1101	DATA
1111	1111	1111	1110	DATA
1111	1111	1111	1111	DATA

Fig. 36.2

change the data stored in ROM. The second main type of memory is the Random Access Memory, RAM. Data is stored in RAM temporarily, and may be changed by the CPU. The CPU can thus READ the data stored in RAM as well as WRITE new data into it. Unless a back-up battery is used, the data stored in RAM will be wiped out when the power supply is turned off. This is why such memory is called **volatile**. ROM, on the other hand, retains its data even when the supply is off. Such memory is called **non-volatile**.

The memory chip

The memory chip consists of a number of memory *locations* where data (information) in the form of digital bits may be stored. Each location can normally store an 8-bit binary number. Each location has a unique 16-bit *address*, as shown in Fig. 36.2. The address may be considered as an envelope. The data is the actual information, the letter inside the envelope.

Microcomputer architecture

Figure 36.3 shows the architecture or organisation of a microcomputer system. It shows the manner in which the various elements are connected to each other in a typical microprocessor computer system. The microprocessor CPU is a single chip containing all necessary circuitry to interpret and execute program instructions in terms of data manipulation, logic and arithmetic operations, and timing and control of the system.

Microprocessors are normally mounted as a 40-pin dual-in-line (DIL) package, as shown in Fig. 36.4.

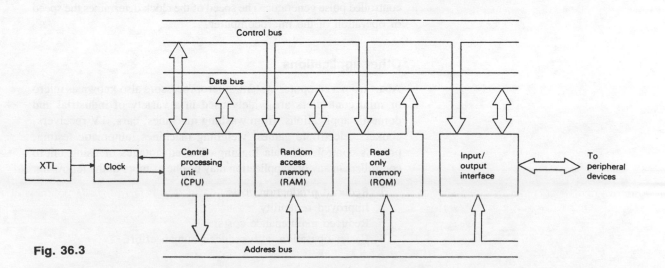

Fig. 36.3

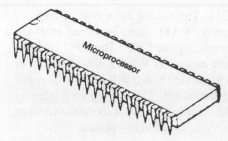

Fig. 36.4

The input/output (I/O) interface

The I/O interface connects the microcomputer to the external devices known as peripherals. It acts as an input or output route for transferring data to and from the microprocessor and peripheral devices such as keyboards, display devices, transducers or drive circuitry for electromechanical devices (stepper motors, relays etc.)

The buses

The main hardware elements in a microcomputer (described above) are interconnected with each other by what is known as a bus structure. A bus is a group of connecting wires or tracks used as paths for digital information that have a common function. There are three main buses in a microprocessor system: the data bus, the address bus and the control bus. The *data bus* is used to transfer data between the microprocessor and the other elements in the system and is normally a bi-directional 8-bit bus.

The *address bus* is used to send the address of memory locations either to retrieve (READ) data from RAM or ROM or to store (WRITE) data into addressed locations in RAM. It is also used to address the input/output interface that may be involved in data transfer. The address bus is uni-directional, carrying 16 bits of digital information simultaneously. The *control bus* is used to send control signals, e.g. clock, read, write etc., from the microprocessor to the other elements of the system. The number of control lines on the bus depends on the microprocessor used and the design of the system.

The clock

A master clock is used in a microcomputer to synchronise the movement of data. This is achieved by using a high-speed crystal-controlled pulse generator. The speed of the clock determines the speed of operation of the microcomputer.

Other applications

Apart from a computer system, microprocessors also known as micro or mini controllers are widely used in a variety of industrial and domestic applications, from washing machines, cars, TV receivers, cookers, electronic games, weighing machines, automatic testing, process control and data logging. The advantages of using micro controllers in any one application may be one or more of the following:

1. Reduced production costs
2. Improved reliability
3. Reduced maintenance costs
4. Easier operation with reduced human effort.

37 Measuring instruments

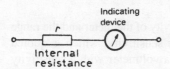

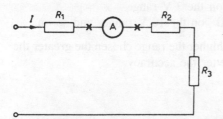

Fig. 37.1 The basic meter

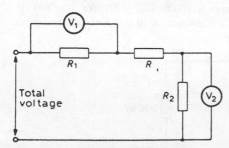

Fig. 37.2 Ammeter A connected to measure current *I*

Fig. 37.3 Voltage measurement

Every meter has certain limitations which have to be taken into account in order to use it in the correct way. The introduction of a measuring instrument into a circuit interferes with its normal condition. The first rule about the use of meters is that this interference is kept to very small proportions and made negligible. A measuring instrument has a resistance known as the internal resistance which can be represented as shown in Fig. 37.1.

Measurement of current

To measure the current in a circuit, the circuit is opened at a convenient place and an ammeter A is inserted in *series* so that the whole current under test passes through it, as shown in Fig. 37.2. The ammeter could be connected anywhere in the circuit provided that the whole current to be measured goes through the meter.

Good quality ammeters have small internal resistances presenting the minimum interference to the flow of current. Ammeters with large internal resistances give inaccurate readings.

Measurement of voltage

Voltage or potential difference exists across two points in a circuit. To measure voltage, a voltmeter is connected *across* the points, e.g. across a resistor *without* interrupting the circuit. Figure 37.3 shows voltmeter V_1 measuring the voltage across R_1 and voltmeter V_2 measuring the voltage across R_2.

Loading effect

As can be seen from Fig. 37.4 the internal resistance of the voltmeter shunts the circuit resistance R as soon as the voltmeter is connected. Part of the current normally passing through R is diverted through the voltmeter. In other words, resistor R now shunted by the internal resistance of the meter presents a smaller effective resistance to the current I. This is known as the loading effect of the voltmeter. To

233

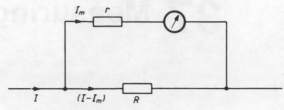

Fig. 37.4 Loading effect I_m is the current diverted into the meter

minimise this effect, the internal resistance of the voltmeter is made as large as possible, over 20 times greater than the load resistance. In this way the shunting effect of the meter resistance is reduced to negligible proportions.

Internal resistance

Internal resistance depends on the sensitivity of the meter and the range chosen and can be calculated from the sensitivity, which is quoted in ohms per volt (Ω/V). For example, a voltmeter with a sensitivity of 1000 Ω/V has an internal resistance of

$$1000 \times 1 = 1000 \ \Omega \quad \text{on the 1 V range}$$
$$1000 \times 3 = 3000 \ \Omega \quad \text{on the 3 V range}$$
$$1000 \times 10 = 10\ 000 \ \Omega \quad \text{on the 10 V range, and so on.}$$

For a given sensitivity, the higher the range chosen the greater the internal resistance and the better the accuracy.

EXAMPLE 1

V_1 and V_2 in Fig. 37.5 are two identical voltmeters having a sensitivity of 20 000 Ω/V. Which meter gives the more accurate reading if both are used on the 10 V range?

Solution

The internal resistance of each meter is 20 000 $\times$ 10 = 200 000 Ω or 200 kΩ. Meter V_1 shunts R_2 which is a resistance of 10 kΩ. Since the resistance

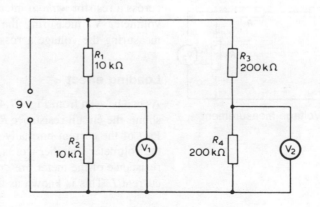

Fig. 37.5

of the meter is much larger than R_2 (20 times greater), it will give an accurate reading (of 4.5 V). On the other hand, meter V_2 shunts a resistor R_4 of equal resistance, namely 200 kΩ. Thus the loading effect of V_2 is large, giving a misleading reading (of 3 V).

Analogue and digital instruments

Analogue instruments such as moving coil meters and oscilloscopes give a continuous indication of voltage, current, etc. *Digital instruments*, on the other hand, provide a reading in discrete steps. They give a direct reading independent of human error, have no moving parts and are smaller in size and cheaper compared with the analogue type.

Types of measuring instruments

Moving coil

The moving coil meter indicates the average, i.e. d.c., value. It may be used for a.c. measurement by employing a rectifier at the input. Sensitivity is in the region of 20 kΩ/V for d.c. and 600 Ω/V for a.c., with a frequency range of up to 2 kHz or more.

Electronic voltmeter (EVM)

This is basically a moving coil meter with an amplifier at its input. A sensitivity in MΩ/V is produced for both d.c. and a.c. with a frequency range of 3 MHz or more.

Digital voltmeter (DVM)

The DVM has a very high sensitivity (in MΩ/V) with a very wide frequency range (above 2 MHz).

The cathode ray oscilloscope (CRO)

Apart from displaying a waveform, the CRO may be used to measure a variety of electrical quantities including voltage (average and peak), periodic time, phase difference and delay. It has a sensitivity and frequency range similar to the EVM and DVM and an input impedance of 1 MΩ.

The multimeter

Although we have only considered types of voltmeters, most commercial instruments combine a few functions into one instrument, known as a *multimeter*. This incorporates a switching arrangement enabling the instrument to be used as an ammeter, a voltmeter and an ohmmeter. It may be an analogue (moving coil) or a digital type.

The oscilloscope

The cathode ray oscilloscope (CRO) may be used to measure frequency. This can be done directly by using the calibrated timebase to measure the periodic time of the displayed waveform. The frequency is then calculated by using the formula $f = 1/t$. This method may be used for a sine wave as well as any other waveform.

A more accurate method of determining frequency of a sine wave is to compare its frequency with a known frequency source. This method consists of switching off the internal timebase of the CRO and applying the known reference frequency to one of the deflecting plates and the unknown frequency to the other plate. Patterns known as Lissajous figures are then displayed, as shown in Fig. 37.6. The unknown frequency may then be determined by counting the peaks of the display. If the unknown frequency, f_Y, is fed to the Y-plates and the known frequency, f_X to the X-plates, then, in cases where only horizontal peaks are produced, as in Fig. 37.6,

unknown frequency f_Y = known frequency f_X × no. of peaks.

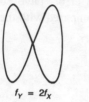

$f_Y = 2f_X$ $f_Y = 3f_X$

Fig. 37.6

Logic state testing instruments

The instruments considered above measure an analogue quantity. To test the logic state of a test point, a *logic probe* must be used (Fig. 37.7). When placed at a test point (or node) the logic probe indicates the logic state of the node, '1', '0' or open circuit. The indication is usually given by one or two LEDs. To change the logic state of a node, a *logic pulser* is used. When placed at a node, the pulser changes the logic state of the node to its opposite. If the node is at logic '1' it changes to '0' and vice versa. The pulser is normally used together with a logic probe to test gates, counters, flip-flops and other digital devices.

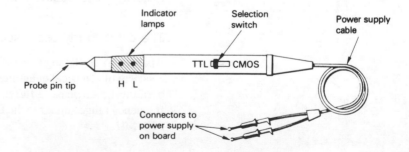

Fig. 37.7

Another very useful logic testing instrument is the *current tracer*. When placed in the proximity of a printed circuit track, the current tracer can indicate the presence of a pulsating current in the track. No electrical contact is necessary. The tracer is used in conjunction with a logic pulser to identify short circuits between a track or a pin on the one hand, and earth or supply on the other. It may also be used to pinpoint shorts between tracks or pins.

Logic and signature analysers

The logic probe and other logic state instruments are of very limited use when testing a microprocessor system. In a bus-structured system the logic level of an individual bus line does not provide adequate information to test the system. To do this it is necessary to look at the logic levels of all the lines of an address or data bus simultaneously. This is carried out by the multi-channel logic analyser, which can monitor a number of inputs simultaneously. Alternatively, the operation of a microprocessor system may be tested by monitoring the sequence of bits appearing on a single point and comparing it with a good, known system. This is carried out by the single input signature analyser.

38 Logical fault finding

There are two types of testing procedures that may be used to diagnose a faulty electronic system, device or printed circuit board (PCB): functional and in-circuit testing. Functional testing inspects the unit-under-test to verify its operation, while in-circuit testing inspects each component on the unit-under-test individually to verify its value, polarity etc. Normally, both methods are employed in sequence. With the development of automatic test equipment (ATE), very fast in-circuit testing, where every component on a PCB, including transistors, gates and counters, is tested separately, has become a possibility. Functional testing also acquired new dimensions with the incorporation of computer processing and control. The principles of logical fault finding, however, are the same, whether testing is carried out manually or automatically.

Fault finding must follow a logical sequence, the purpose of which is to identify the cause of the fault and then to rectify it. The number of tests carried out should be kept to a minimum, and unnecessary or pointless tests must be avoided. Before any test is carried out, visual inspection of the circuit and the various components should be made to check for burnt-out components, obvious breaks in the printed circuit, and so on. Such inspection should take no more than two or three minutes and with experience will become instinctive. This having been done, without success, proper fault finding procedure can then begin.

First, a **functional test** is carried out where the operation of the board is tested and an attempt is made to identify the faulty block and the suspected faulty component. Before the faulty component is replaced, an **in-circuit measurement** of the component is carried out to confirm the fault.

Functional tests

This type may be divided into two series of tests.

Series I known as **dynamic tests** are applied to the complete electronic equipment in order to isolate the faulty stage or block. Once the fault has been traced to a specific block, then *Series II* or **static** tests are applied to pin-point the cause of the fault to one or two possible faulty components such as a resistor, a capacitor, etc.

Dynamic tests

These are the first set of tests that are carried out on the faulty electronic equipment. The tests must begin from the output going towards the input. Using the **half-split method** the equipment is initially divided into two appropriate sections. The section at the output is tested first by **injecting** a signal similar to the one normally present at that point. If normal output is produced, the fault must then lie in the remaining section. This section is now divided into two appropriate sub-sections and the same procedure is repeated, and so on until the fault is isolated to the smallest distinguishable stage such as the output stage, i.f. or video amplifier, divider, decoder or a single gate.

EXAMPLE 1: A radio receiver (Fig. 38.1)

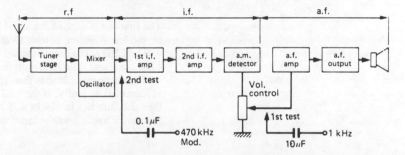

Fig. 38.1

The most appropriate division is the a.f. on one hand and the i.f./r.f. on the other. The a.f. section is first tested by injecting a 1 kHz signal at its input (the volume control), via a coupling capacitor (10–50 μF). Low, distorted or no output indicates a faulty a.f. section. The a.f. section is now divided into two sub-sections: the output stage and the pre-amp. Each sub-section is then tested starting from the output. If the fault does not lie in the a.f. section, a pure 1 kHz tone should be heard from the loudspeaker. The fault must then be in the i.f./r.f. section.

A quick check on whether the fault is in the a.f. section or not is the **screwdriver** test. Touch the input of the a.f. section (with the volume control on full) with a screw-driver. A noise from the loudspeaker should be heard if the a.f. section is sound.

If it is found that the fault lies in the i.f./r.f. section a further division is made, namely an r.f. section and an i.f. section. The i.f. section is tested first by injecting an amplitude-modulated 470 kHz signal via a coupling capacitor (0.01–0.1 μF) into the input of the i.f. section, e.g. into the base of the first i.f. transistor. For f.m. receivers a frequency-modulated 10.7 MHz test signal should be used. A sound i.f. section will produce a pure tone (400–600 Hz) at the loudspeaker. Otherwise further sub-divisions are made until the faulty stage is discovered, e.g. one i.f. amplifier or the detector.

If the fault is in the r.f. section, this section is sub-divided if possible and tested as follows. An amplitude-modulated 1000 kHz signal is injected via a coupling capacitor (0.01–0.1 μF) into the input of the stage. The receiver should then be tuned to receive the radio frequency of 1000 kHz or 300 m on the MW. Another appropriate frequency must be used for f.m. receivers.

An alternative to signal injection is the **signal tracing** method where a radio receiver or other electronic equipment is switched on and tuned to a station in the case of a receiver or fed with normal input signal. The presence or otherwise of a signal at appropriate points is confirmed starting from the output. The signal is observed on a cathode ray oscilloscope to test for its waveshape and amplitude.

The above principles for dynamic tests can be equally applied to any faulty electronic equipment using appropriate divisions and signal frequencies.

EXAMPLE 2: A digital divider and display (Fig. 38.2)

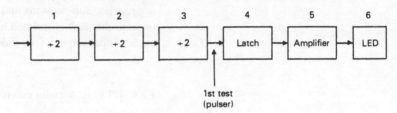

Fig. 38.2

The first test is applied at a suitable point 'half-way' along the divider chain, as shown. A pulser is used to change the logic state at the input of block 4. The LED at the output should change state if the latch, amplifier and the LED are sound. The fault must then be in one of the dividers preceding block 4. The same procedure is then repeated, using the pulser until one divider is identified as faulty. If the LED does not change state after the first test, then the fault lies in blocks 4, 5 or 6. The same procedure is repeated with the pulser applied to the input of the amplifier and so on.

Principles of static tests

Having identified the faulty stage, this series of tests is used to identify the faulty component.

1. Start with d.c. tests. Use a meter having a sensitivity of at least 20 000 Ω/V.
2. Take voltage measurements only. If current value is required, estimate it by measuring the voltage across a known resistor.
3. If the fault is not revealed by d.c. measurements, then, and only then, apply signal tests to the stage.

Application to single stage amplifier (refer to Fig. 38.3)

Usually the normal d.c. readings for the stage are given. If not they could be estimated to a reasonable degree of accuracy. The voltage readings of the faulty stage are then compared with the normal readings. This comparison should then give a good indication of where the fault lies. First determine the static state of the transistor:

(a) transistor at cut-off producing no output signal (or going towards cut-off) or

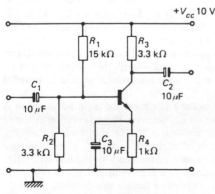

Fig. 38.3 Normal readings: e 1.1 V, b 1.72 V, c 6.37 V

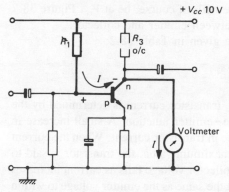

Fig. 38.4 R_3 o/c, transistor cut off: e 0.3 V, b 0.94 V, c 0.3 V

Table 38.1 Cut-off condition

Fault		Cause
(1) e	zero	
b	zero	
c	V_{cc}	R_1 o/c
V_{be}	zero	
(2) e	high	
b	normal	
c	V_{cc}	R_4 o/c
V_{be}	low	
(3) e	low ($\simeq V_c$)	
b	low	
c	low ($\simeq V_e$)	R_3 o/c
V_{be}	normal	

(b) transistor at saturation producing low distorted output or going towards saturation or

(c) transistor at normal static state.

Once the state of the transistor has been established, the next step is to work out the cause for cut-off or saturation. (When functioning under normal static state, the fault is an a.c. fault which will be dealt with later.)

Cut-off condition

Transistor current ceases to flow, i.e. cut-off, if (a) b—e forward bias is zero or (b) the current path is interrupted, namely when R_3 is open circuit (o/c), when R_4 is o/c, or the transistor itself is faulty. Normally when the transistor is at cut-off, the collector goes up to d.c. supply V_{cc}. However, if R_3 goes o/c the collector is 'floating' and theoretically it is at base potential. As soon as a voltmeter is connected to measure the collector voltage, the base—collector junction becomes forward bias as shown in Fig. 38.4. Current will flow through the meter, the b—c junction and R_1 giving a small reading on the meter. This reading is entirely due to the meter resistance.

Similarly, when cut-off is caused by R_4 going o/c, the emitter is 'floating' and theoretically should be at base potential. However, a voltmeter connected to measure that voltage provides a path for current to flow and forward biases the base—emitter junction, giving a reading on the meter slightly higher than normal (Fig. 38.5).

Table 38.1 gives a summary of the above faults. Note that in the table a high V_{be} refers to an increase of between 0.1 to 0.2 V in the base—emitter forward bias.

Transistor faults also produce cut-off conditions. Voltage readings in such cases depend on the nature of the fault as well as the component values. For example, a b—e short circuit (Fig. 38.6) cuts the transistor off and puts R_4 in parallel with R_2 causing base and emitter voltage to go down to a low value determined by the potential divider

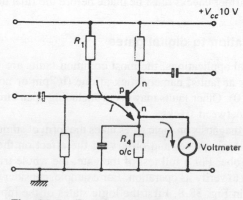

Fig. 38.5 R_4 o/c, transistor cut off: e 1.25 V, b 1.74 V, c 10 V

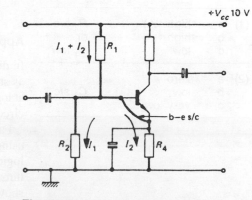

Fig. 38.6 b—e s/c, transistor cut off: e 0.48 V, b 0.48 V, c 10 V

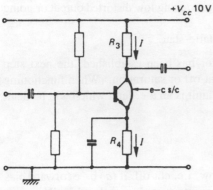

Fig. 38.7 e–c s/c, transistor cut off: e 2.29 V, b 1.77 V, c 2.99 V

Table 38.2

Fault			Cause
(1)	e	zero	
	b	normal	
	c	V_{cc}	
	V_{be}	very high, cannot be sustained by functioning junction	b–e o/c
(2)	e	low	
	b	low	
	c	V_{cc}	b–c o/c
	V_{be}	normal	

Table 38.3

Fault			Cause
(1)	e	high (V_c)	R_2 o/c
	b	high	or
	c	low	R_1 low
(2)	e	zero	
	b	low	C_3 s/c
	c	very low	

$R_1 - R_2/R_4$. Collector voltage will, of course, be at V_{cc}. Figure 38.7 shows the effect of a s/c between emitter and collector.

Other transistor faults are given in Table 38.2.

Saturation condition

As explained in Chapter 21, transistor current is determined by the forward bias across the base–emitter junction. A small increase in this bias produces a large rise in transistor current. When the current through the transistor is at maximum value, the transistor is said to be saturated or bottomed. Collector voltage falls as current increases until at saturation it is almost the same as the emitter voltage to within 0.1–0.5 V. In general at saturation, the emitter, base and collector potentials are approximately at the same level. Table 38.3 summarises.

Normal static state

Normal d.c. voltages with no or low output indicate an a.c. fault, e.g. an open circuit coupling capacitor. Before replacing a suspected o/c capacitor, place a good capacitor of similar value across the suspected component to confirm the fault. An open circuit decoupling capacitor (C_3 in Fig. 38.3) produces low undistorted output. A leaky or short circuit coupling capacitor usually introduces a change in the d.c. state of the transistor depending on the preceding or following stages.

In fault finding the following should be remembered:

1. Do not jump to conclusions by merely comparing one reading with its normal value. The whole set of readings (in the case of a transistor stage the emitter, base and collector) should be noted on paper and compared with the normal set of readings.
2. If accurate readings (up to 0.01 V are possible on a 20 000 Ω/V meter) are taken, then in the vast majority of cases two equal readings indicate a s/c. However, this is not always true and further checks must be made before the final decision is made.

Application to digital gates

In digital applications, the most common faults are what are known as stuck-at faults, namely stick-at-one (IC pin or node permanently at logic 0). Other faults may occur including open circuit pins or short circuit between tracks.

Fault diagnosis on logic gates takes the form of stimulating the inputs using a logic pulser and observing the effect on the output using a logic probe. For a full test of the gate, the whole truth table is gone through to verify its operation. For example, consider the digital circuit shown in Fig. 38.8. First the logic states of the input and output of each gate are noted and checked against their respective truth tables.

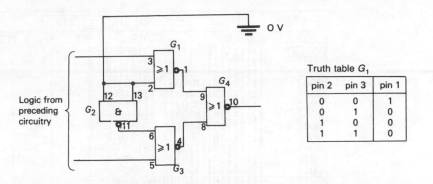

Fig. 38.8

If a gate is then suspected, then a further test on the gate is carried out using a logic pulser and probe. Consider gate G_1. Input pin 2 is permanently at logic 0. To test the gate, a pulser is placed at input pin 3 and a logic probe is placed at output pin 1. By referring to the truth table for NOR gate, it can be seen that with one input (pin 2) at logic 0, a change in the output is produced if the other input (pin 3) is made to change its logic state. For instance, if pin 3 was originally at logic 0, then output pin 1 will be at logic 1. If the pulser is then used to change the state of pin 3 to logic 1, then provided that the gate is sound, a change will be observed by the probe at output pin 1 from logic 1 to logic 0. An opposite change will be observed if pin 3 was originally at logic 1. Similar tests may be applied to the other gates. In each case, attention must be paid to the truth table of the particular gate to ensure that the application of the pulser to one of its inputs does, in fact, produce a change in the output, otherwise the test will be inconclusive.

Application to microprocessor systems

Fault diagnosis on bus-structured microprocessor systems takes the form of acquiring the sequence of addresses and data that appear on the address and data buses and comparing it with a known-good list. A fault, such as a data bus line 3, D_3 stuck-at-0 will be indicated by a permanent zero at D_3. Such a listing, known as **state listing**, is produced by a logic analyser. A typical state listing display is shown in Fig. 38.9.

Alternatively, a signature analyser may be used to capture the sequence of bit stream at a node, known as a signature, and compare it with a correct signature. A difference indicates a fault.

Fault finding examples

In each example separate fault conditions are described. The fault in each case is assumed to be the result of a failure in a single component, but there may be more than one component that could cause the fault.

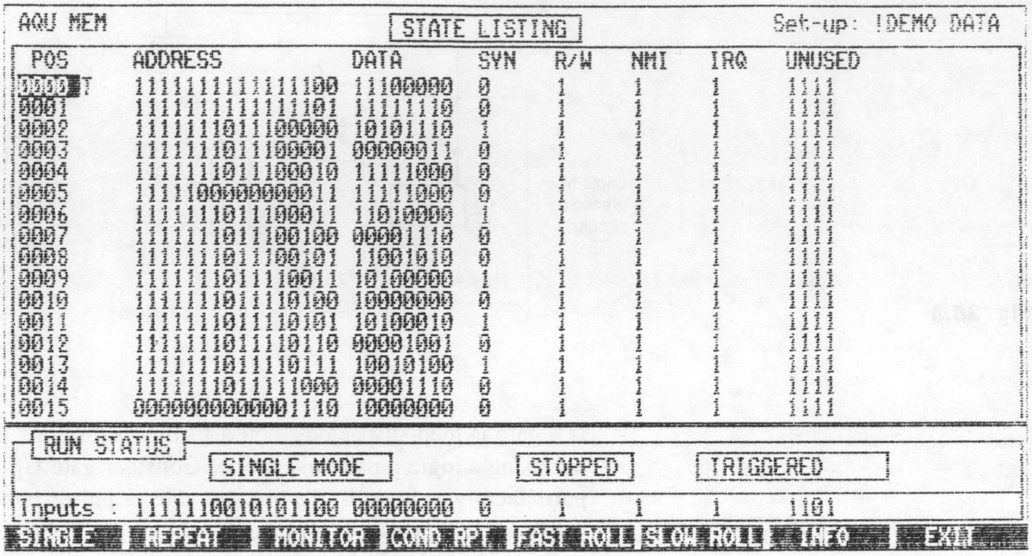

```
 AQU MEM                        STATE LISTING              Set-up: !DEMO DATA
  POS      ADDRESS              DATA       SYN   R/W   NMI    IRQ    UNUSED
 0000  T   1111111111111100     11100000    0     1     1      1     1111
 0001      1111111111111101     11111110    0     1     1      1     1111
 0002      1111111011100000     10101110    1     1     1      1     1111
 0003      1111111011100001     00000011    0     1     1      1     1111
 0004      1111111011100010     11111000    0     1     1      1     1111
 0005      1111100000000011     11111000    0     1     1      1     1111
 0006      1111111011100011     11010000    1     1     1      1     1111
 0007      1111111011100100     00001110    0     1     1      1     1111
 0008      1111111011100101     11001010    0     1     1      1     1111
 0009      1111111011110011     10100000    0     1     1      1     1111
 0010      1111111011110100     10000000    0     1     1      2     1111
 0011      1111111011110101     10100010    1     1     1      1     1111
 0012      1111111011110110     00001001    0     1     1      1     1111
 0013      1111111011110111     10010100    1     1     1      1     1111
 0014      1111111011111000     00001110    0     1     1      1     1111
 0015      0000000000001110     10000000    0     1     1      1     1111

  RUN STATUS
            SINGLE MODE                STOPPED          TRIGGERED

 Inputs :  1111110010101100 00000000    0     1     1      1     1101

 SINGLE   | REPEAT   | MONITOR |COND RPT |FAST ROLL|SLOW ROLL|  INFO   |  EXIT
```

Fig. 38.9

All test potentials are measured in volts with respect to the chassis, using a 20 000 Ω/V voltmeter.

1. D.C. coupled amplifier

(Refer to Fig. 38.10)

Test point	1	2	3	4
Normal readings	0.7	0.1	5.1	4.5
Fault A (no output)	0	0	8.65	8
Fault B (no output)	0.01	0.01	8.4	7.75
Fault C (no output)	0.48	0	0.02	0
Fault D (no output)	0.7	0.1	5.1	4.7

FAULT A

TR_1 is cut off, emitter is zero, and collector is at V_{cc}. Cut-off caused by TR_1 base (test point 1) having zero volts. Since the two transistors are d.c. coupled, TR_2 base voltage increases with collector of TR_1, causing TR_2 to saturate.

Answer: R_1 o/c

FAULT B

TR_1 is cut off. Voltages at test points 1 and 2 are equal suggesting a short circuit. This voltage is determined by resistor chain R_1 (1.6 MΩ) in series with R_2 (70 kΩ) with the latter shunted by R_4 (1 kΩ). As in Fault A, TR_2 is saturated.

Answer: TR_1 b–e s/c.

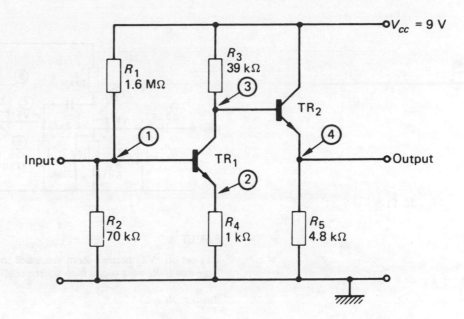

Fig. 38.10

FAULT C

TR_1 and TR_2 are both cut off. TR_2 emitter voltage is at zero (test point 4) indicating cut off. TR_1 has low collector voltage suggesting saturation, while its emitter (test point 2) is zero. Hence TR_1 must be at cut off caused by the collector resistor going open circuit. TR_1 base is down because of a large base current flowing through high resistor R_1 (1.6 MΩ).

Answer: R_3 o/c.

FAULT D

TR_1 is normal. A saturated TR_2 would produce a high emitter voltage. Since the emitter is almost normal and with the absence of any possible a.c. fault, TR_2 must be at cut off. A faulty transistor, e.g. b–c o/c, would produce zero emitter voltage. An open circuit b–e junction would produce low emitter voltage. The fault must be R_5 o/c with emitter reading (test point 5) due to meter resistance.

Answer: R_5 o/c.

2.   Two-stage R–C coupled fet amplifier

(Refer to Fig. 38.11)

Test point	1	2	3	4	5	6
Normal readings	15	2.4	5.55	0	3.3	12.75
Fault A (no output)	15	0	0	0	3.3	12.75
Fault B (no output)	15	2.4	5.55	5.55	7.5	9.9
Fault C (no output)	15	2.4	5.55	0	4.1	15
Fault D (no output)	15	2.4	5.55	0	6.1	10.8
Fault E (low output)	15	2.4	5.55	0	3.3	12.75

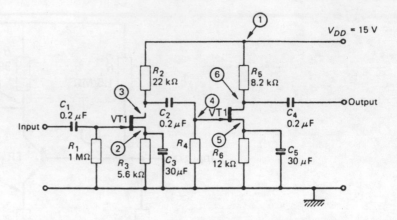

Fig. 38.11

FAULT A

VT_1 cut off. VT_2 normal. Zero drain and source voltages suggests R_2 o/c. Note that if R_3 were o/c, a high source reading would have been obtained.

Answer: R_2 o/c.

FAULT B

VT_1 normal. VT_2 saturated (source high, drain low). A fet may saturate if the gate–source reverse bias V_{GS} is reduced. With an n-channel fet used in the circuit, this may be produced either by the gate voltage going positive (towards the source) reducing V_{GS} or by removing the gate leakage resistor (also see Fault D). Test points 3 and 4 are equal suggesting C_2 short circuit. This will increase VT_2 gate voltage which reduces the reverse bias and increases the current through the transistor.

Answer: C_2 s/c.

FAULT C

VT_1 normal. VT_2 drain at 15 V suggests cut-off. The increase in source voltage (test point 5) points to R_6 o/c with the voltage reading at the source due to meter resistance. Note that a faulty fet such as an open circuit junction would produce cut-off but would give a zero source reading.

Answser: R_6 o/c.

FAULT D

VT_1 is normal. VT_2 is saturated (high source, low drain). Since gate voltage remains at zero, saturation is due to gate leakage resistor going open circuit, thus removing the gate–source reverse bias and increasing the drain current.

Answer: R_4 o/c.

FAULT E

All d.c. voltages are normal. Hence fault is a.c. Low output suggests an open circuit decoupling capacitor introducing negative feedback and reducing the gain of the amplifier.

Answer: C_3 or C_5 o/c.

3.   Stabilised power supply

(Refer to Fig. 38.12)

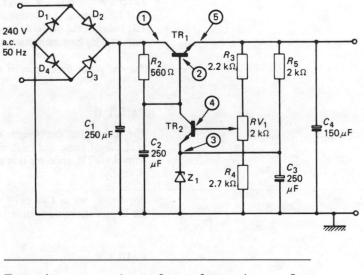

Fig. 38.12

Test point	1	2	3	4	5
Normal readings	18	14.1	8.75	9.35	13.3
Fault A	19.5	2.65	0	0.7	2.0
Fault B	19.8	13.5	8.75	9.35	13.5
Fault C	20	20	8.65	0	19.3
Fault D	19.5	0	0	0	0
Fault E	20	20	0	0	0

Note   Always start by noting the zener reference voltage. A low or zero reading indicates that the zener is not at breakdown either because of a faulty zener or a fault in the resistor feeding it (R_5 in Fig. 38.12).

FAULT A

Zener reference voltage is zero, which points to a short circuit between test point 3 and chassis. Note that an open circuit R_5 would not give zero reading at test point 3 since the meter resistance would complete the emitter circuit of TR_2 giving a voltage reading.

Answer: Z_1 or C_3 s/c.

FAULT B

Zener voltage is normal. Test points 2 and 5 are the same potential suggesting a short circuit. TR_1 is cut off, hence its collector (test point 1) has increased.

Answer: TR_1 b−e s/c.

FAULT C

Zener voltage is 0.1 V lower than normal indicating that, while the zener is at breakdown, it is taking a smaller current than normal. This is due to TR_2 being at cut-off (note its collector at 20 V). The cut-off is due to zero volts at TR_2 base indicating a break in the bias chain.

Answer: R_3 or upper part of RV_1 o/c.

FAULT D

The zero volt at the zener is caused by the zero volts at test point 5. TR_2 is at cut-off (base and emitter are both zero volts). Test point 1 is almost normal yet TR_2 collector is at zero volts. This indicates a fault in components R_2 or C_2.

Answer: R_2 o/c or C_2 s/c. (C_2 s/c would anyway cause R_2 to go o/c due to the large current through it.)

FAULT E

Zero zener voltage is caused by zero voltage at test point 5. Voltage across b−e junction of TR_1 = test point 2 − test point 5 = 20 − 0 = 20 V. A pn junction cannot sustain a forward bias higher than 0.7 V.

Answer: TR_1 b−e o/c.

4. Output stage

(Refer to Fig. 38.13)

Test point	1	2	3	4	5
Normal readings	0.61	6.3	6.7	6.45	12
Fault A (no output)	0	12	12	12	12
Fault B (no output)	0.19	0	0.79	1.09	12
Fault C (very low output)	0.4	1.1	1.08	1.08	12
Fault D (no output)	0.61	5.85	6.22	6.0	12

Note   Start by looking at test point 4 where the normal reading should be approximately $\frac{1}{2} V_{cc}$. If test point 4 is high, then TR_3 is conducting more than TR_2 and vice versa.

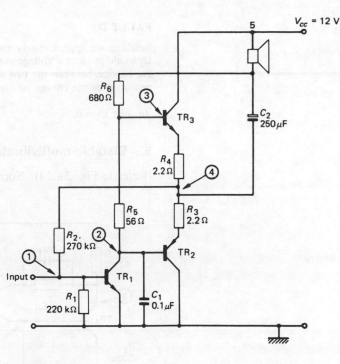

Fig. 38.13

FAULT A

Test point 4 is at V_{cc}. TR_3 is saturated and TR_2 is cut off. TR_1 collector (test point 2) is at V_{cc} indicating TR_1 cut-off. TR_1 base is zero volts, hence fault is in bias chain.

Answer: R_2 o/c.

FAULT B

Test point 2 is zero volts suggesting a short circuit to chassis. Such a short circuit would cause TR_2 b–e junction to conduct heavily, charging C_2 and reducing the potential at test point 4. Low TR_1 base voltage is due to low voltage at test point 4.

Answer: TR_2 b–c s/c or C_1 s/c.

FAULT C

Test point 4 is low. C_2 is charged by TR_2 which is conducting more than normal due to its low base voltage (test point 2). The low potential at test point 2 is not due to a saturated TR_1. Note its low b–e forward bias V_{be} of 0.4 V instead of the normal 0.61 V. TR_1 must thus be at cut-off caused by open circuit collector resistance. TR_1 is not fully cut off. It conducts through the b–e junction of TR_2 thus charging C_2 and bringing test point 4 down to 1.08 V.

TR_3 base is at the same potential as TR_1 collector since no current flows through R_5. TR_3 is at cut-off.

Answer: R_6 o/c.

FAULT D

Readings are approximately normal. Hence the fault is a.c. An open circuit C_2 would produce a voltage at test point 4 slightly lower than normal, since the balance between the two output transistors TR_2 and TR_3 is no longer produced by the charge on capacitor C_2.

Answer: C_2 o/c.

5. Bistable multivibrator

(Refer to Fig. 38.14). Normal readings are as follows.

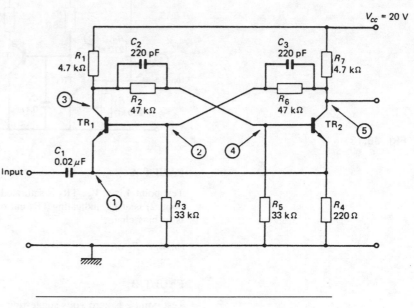

Fig. 38.14

Test point	1	2	3	4	5
State I (TR_1 off)	1.1	0.45	18.5	1.3	1.2
State II (TR_1 on)	1.1	1.3	1.2	0.45	18.5

In each of the following cases the output fails to change when an input pulse is received.

Test point	1	2	3	4	5
Fault A	0.95	0.95	18.5	1.6	1.0
Fault B	0.95	0	18.5	1.6	1.0
Fault C	0.1	0.7	0.1	0.05	18.5
Fault D	1.1	0.45	18.5	1.3	1.2
Fault E	7.7	8.0	19.0	8.0	19.0

Note The bistable has two normal states in which it may settle, hence two normal sets of readings. When locating a fault, first determine the state of each transistor, then compare the faulty readings with the normal readings to determine the reason for the failure to change over from one state to another.

FAULT A

TR_1 off, TR_2 on. TR_1 b–e voltage is zero suggesting a short circuit between base and emitter. Voltage at test point 1 (0.95 V) is now reduced due to the shunting effect of R_4 upon R_3.

Answer: TR_1 b–e s/c.

FAULT B

TR_1 off, TR_2 on. TR_1 base voltage is zero suggesting a faulty bias chain $R_6 - R_3$.

Answer: R_6 o/c.

FAULT C

TR_2 is cut off. TR_1 seems saturated; however the low emitter voltage (test point 1 at 0.1 V) indicates cut-off due to open circuit collector resistor R_1. TR_1 collector voltage reading (test point 3) is due to meter resistance. Emitter voltage reading (test point 1) is due to a large base current flowing through forward biased b–e junction of TR_1.

Answer: R_1 o/c.

FAULT D

D.C. readings are normal. Hence fault is a.c.

Answer: C_1 o/c.

FAULT E

The readings are symmetrical with both TR_1 and TR_2 at cut-off. Hence the fault must be common to both transistors. R_4 is the only common component that affects the d.c. conditions of the circuit. Reading at test point 1 is due to meter resistance.

Answer: R_4 o/c.

6. Digital circuit

(Refer to Fig. 38.15)

Circuit description: the circuit in Fig. 38.15 shows part of a traffic light controller. Pulses applied at the input are counted by the binary counter IC5, which divides the input by a factor of 5 as shown in the output functions table. The three outputs of IC5, namely C (pin 8), B (pin 9) and A (pin 12) form a binary code for each function of the traffic light system. For instance, when IC5 output CBA are

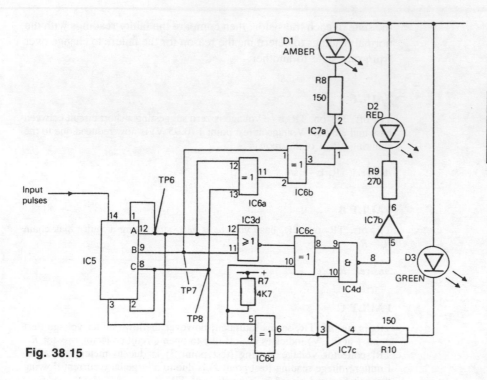

Fig. 38.15

Output functions of IC5

C	B	A	Function
0	0	0	Amber
0	0	1	Red
0	1	0	Red
0	1	1	Red/amber
1	0	0	Green

000 then the Amber light is on, and the Green and Red are off, and for an output of 011 then both Amber and Red are on and the Green is off, and so on. The counter resets itself to 000 after Green. The combinational gates following the counter form a steering circuitry which ensures the correct LED or LEDs are activated. IC7a, b and c are driver amplifiers.

From the following symptom and readings, state the possible fault.

SYMPTOM: NO GREEN

The following logic states were observed on the respective nodes when the circuit was taken through its function by applying individual pulses at the input using a pulser. All readings are observed using a logic probe.

	IC5			IC6d			IC7c	
Pins	C	B	A	4	5	6	3	4
	0	0	0	0	1	1	1	1
	0	0	1	0	1	1	1	1
	0	1	0	0	1	1	1	1
	0	1	1	0	1	1	1	1
	0	0	0	0	1	1	1	1
	0	0	1	0	1	1	1	1

Answer: From the outputs of IC5 it can be seen that the Green sequence (100) is missing, hence IC7c pins 3 and 4 are at logic 1, keeping Green led off.

The fault is output C of IC5 stuck-at-0. The counter thus resets itself to 000 immediately after the Red/Amber sequence.

Suspected fault: IC5 pin 8 or pin 2 stuck-at-0.

COMPONENT TESTING

Testing of components is normally carried out by the use of an ohmmeter. The resistance of the component is measured and compared with that of a sound component.

Resistors and inductors

The resistance of a resistor or an inductor may be measured accurately, thereby giving an indication of a possible fault. Resistors tend to increase their resistance or go open circuit. They very seldom go short circuit. Inductors (and transformers) go open circuit which is easy to detect. They may also develop a shorted turn (i.e. a short circuit between turns) which is fairly common but difficult to detect by ohmmeter measurements, or may develop a short circuit between coil and core which is easy to confirm.

Capacitors

A capacitor may go open circuit, short circuit or leaky. A short circuit or a leak (i.e. a low resistance reading) can easily be detected by ohmmeter measurement. On the other hand it is more difficult to confirm an open-circuit fault. When the ohmmeter terminals are connected across a capacitor, the ohmmeter battery voltage begins to charge up the capacitor. Provided the capacitor has a large value the meter may be observed to kick towards the zero reading and, as the capacitor charges up, its reading slowly rises to infinity. Failure to do that indicates an open circuit. However, a small capacitor would charge up too quickly for the meter to register the movement.

Electrolytic capacitors may be tested for an open circuit because of their high capacitance as well as their normally high leakage current which gives a low resistance reading (few 100 kΩ) when connected in the correct polarity.

Capacitors may change their value, in which case a bridge test where the value of the capacitance is measured is required to detect the fault.

Diodes

The testing of semiconductor devices normally involves the measurement of the forward and reverse resistance of a pn junction. For a diode the ohmmeter is first connected as shown in Fig. 38.16(a) with a negative terminal connected to the anode, and the positive terminal connected to the cathode. The diode is, therefore, reverse biased giving a very high reading (in MΩ). The polarity is then

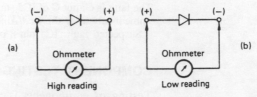

Fig. 38.16 Testing a diode

reversed as shown in Fig. 38.16(b) giving a forward bias to the diode. A low forward resistance will be registered on the meter (600–1000 Ω on a moving coil, and few ohms on an EVM or DVM). A diode giving a low reading in both directions may be assumed to be a short circuit.

The ohmmeter test may also be used to identify the anode and cathode terminals of a diode. When the ohmmeter is reading low resistance (i.e. forward biased), as shown in Fig. 38.16(b), the polarity of the ohmmeter is the same as the polarity of the diode, i.e. the positive (black) terminal is connected to the anode and the negative (red) terminal is connected to the cathode.

Bipolar transistors

The bipolar transistor consists of two pn junctions which are tested individually in the same way as a diode junction. The forward and reverse resistance of the b–e and b–c junctions are measured separately. Normal readings are of the same order as for an ordinary diode. The resistance between the collector and the emitter is also tested and should give a very high reading (in MΩ) or infinity in both directions.

Junction fet

The ohmmeter is used to measure the resistance of both junctions as well as the channel path itself. For a sound fet the following readings should be obtained:

> Drain-gate (forward bias) — low (40 Ω).
> Drain-gate (reverse bias) — infinity (MΩ).
> Gate-source (forward bias) — low (40 Ω).
> Gate-source (reverse bias) — infinity (MΩ).
> Drain-source or channel resistance (in either direction) — low (100 Ω).

The SCR

Both the forward and reverse resistance between the anode and the cathode are very high. Triggering may be obtained by connecting the ohmmeter to forward bias the SCR as shown in Fig. 38.17. If the gate is now shorted to the anode, gate current flows which triggers the SCR and makes it conduct. The ohmmeter reading thus drops to

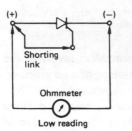

Fig. 38.17 Testing an SCR

a low value (few 100 Ω). The ohmmeter reading remains at that low value even if the shorting link is removed.

In-circuit testing

It is always advisable to check a suspected faulty component while it remains in the circuit. Once the fault is confirmed then a replacement may be made. These tests are carried out with the aid of an ohmmeter and usually referred to as *continuity tests*. In carrying out in-circuit measurements, precautions have to be taken to ensure that the shunting effect of other components is minimised or taken into account.

In-circuit measurement of resistance

Consider the circuit in Fig. 38.18 where R_x would read the total resistance across A and B. Also $R_1 + R_2 = 10$ kΩ + 10 kΩ = 20 kΩ offers a parallel path for the ohmmeter current to flow giving a lower reading on the meter. If R_x is sound, the ohmmeter reading would be

$$\frac{10 \times 20}{10 + 20} = \frac{200}{30} = 6.67 \text{ k}\Omega$$

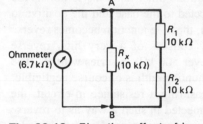

Fig. 38.18 Shunting effect of in-circuit measurement of a resistance

If R_x is open circuit, the ohmmeter would measure the shunting resistance only, namely 20 kΩ.

The circuit in Fig. 38.18 has a parallel path resistance which is comparable to the resistance of the component under test. Hence its effect has to be allowed for. In Fig. 38.18, $R_x = 220$ Ω. The shunting resistance is 10 kΩ + 10 kΩ = 20 kΩ which is 100 times greater than the component under test. In this case the effect of the shunting path may be neglected.

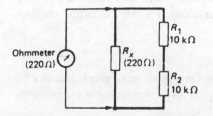

Fig. 38.19 Shunting effect of in-circuit measurement of a resistance

In cases where the suspect component has a value far greater than the total resistance of the shunting path (Fig. 38.20), an in-circuit test becomes unreliable. The ohmmeter gives a reading approximately equal to the total resistance of the shunting path. For the circuit in Fig. 38.20.

$$\text{Meter reading} = \frac{100 \times 20}{100 + 20} = \frac{2000}{120} = 19 \text{ k}\Omega$$

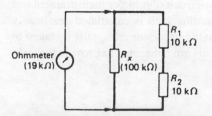

Fig. 38.20 Shunting effect of in-circuit measurement of a resistance

Note that a sound component always registers a reading equal to or smaller (due to the shunting path) than its real value and never a higher reading. If a higher reading *is* registered, then the component has gone either high or open circuit.

In many cases a shunting path may include a pn junction either in the form of a diode or a b—e or a b—c junction of a transistor. Figure 38.21 shows a simple single-stage transistor amplifier. To test resistor R_2 assume that the ohmmeter is connected with its positive terminal to the base and its negative terminal to chassis. The b—e junction becomes forward biased by the ohmmeter battery providing a low

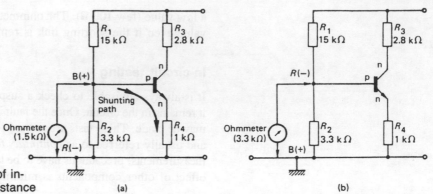

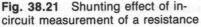

Fig. 38.21 Shunting effect of in-circuit measurement of a resistance

resistance shunting path through R_4 as shown in Fig. 38.21(a) and giving an unreliable reading. If on the other hand the terminals are reversed with the negative connected to the base and the positive to chassis as shown in Fig. 38.21(b), the b–e junction becomes reverse biased. The shunting path now consists of a very high reverse resistance of the b–e junction (over 500 kΩ) in series with R_4. The effect of such a high resistance shunting path is of course negligible.

In general therefore when measuring a resistance in-circuit, the ohmmeter terminals must be connected in such a way as to reverse bias any pn junction that may form part of a parallel path in order to minimise the shunting effect. In practice two separate ohmmeter readings are taken with the ohmmeter terminals connected in both directions. *The higher reading is always the more accurate reading.*

Inductors

Inductors (or transformers) have very small resistances. Hence only a very low resistance shunting path has any noticeable effect on the accuracy of the ohmmeter reading.

Capacitors

Leakage may be confirmed by in-circuit ohmmeter measurement test only if the resistance of the shunting path is calculated reasonably accurately. In testing an electroyltic capacitor care must be taken to ensure that the ohmmeter terminals are connected to provide the right polarity to the capacitor.

Diodes and transistors

In testing a suspected diode or a transistor the forward and reverse resistances of the junctions are measured. In the forward direction the resistance of the junction is low, hence the effect of the shunting path is very small. For example, the forward resistance reading of the b–c junction of the transistor in Fig. 38.21 is approximately

400 Ω. In the reverse direction the resistance is very high with the ohmmeter reading mainly the resistance of the shunting path. For the reverse resistance of the b–c junction of the transistor in Fig. 38.21, the ohmmeter reading will be approximately $R_3 + R_1 = 2.8$ kΩ + 15 kΩ = 17.8 kΩ.

Note that for an open circuit junction, both the forward and reverse readings will be the same (17.8 kΩ for an open circuit b–c junction in Fig. 38.21).

39 Useful Exercises II

CHAPTERS 20–38. For answers see Appendix VII.

1 The reverse resistance of a germanium diode is 2 MΩ at an ambient temperature of 25°C. If the ambient temperature is raised to 70°C the reverse resistance will be approximately
   (*a*) 1 kΩ
   (*b*) 100 kΩ
   (*c*) 2 MΩ
   (*d*) 5 MΩ

2 Refer to Fig. Q.1. If D_1 is a silicon device the expected voltage between points A and B will be
   (*a*) 0 V
   (*b*) 0.3 V
   (*c*) 0.6 V
   (*d*) 5 V

3 Refer to Fig. Q.1. The voltage across R_1 is
   (*a*) 0 V
   (*b*) 0.6 V
   (*c*) 5 V
   (*d*) 10 V

4 Refer to Fig. Q.2. With a mains input of 240 V r.m.s. the peak inverse voltage across each diode will be approximately
   (*a*) 170 V
   (*b*) 240 V
   (*c*) 350 V
   (*d*) 700 V

5 The anode current of an SCR may be cut off by applying
   (*a*) negative voltage to the gate
   (*b*) positive voltage to the gate
   (*c*) negative voltage to the cathode
   (*d*) negative voltage to the anode

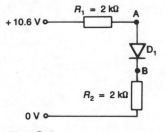

Fig. Q.1

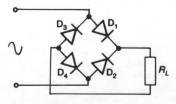

Fig. Q.2

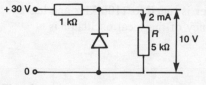

Fig. Q.3

6 Refer to Fig. Q.3. The current taken by the zener is
   (*a*) 20 mA
   (*b*) 18 mA
   (*c*) 2 mA
   (*d*) 0

7 Refer to Fig. Q.3. If the resistance of the load resistor R is increased to 10 kΩ, the current in the zener diode will
   (*a*) decrease by 1 mA
   (*b*) remain constant
   (*c*) increase by 1 mA
   (*d*) increase by 2.0 mA

8 In a series $L-R-C$ circuit, the current will have a maximum value when
   (*a*) the values of L and C are equal
   (*b*) the values of X_L and X_C are equal
   (*c*) X_L is a maximum
   (*d*) R is a maximum

9 A transmission line has a loss of 6 dB. If a signal of 1 V is fed in, the voltage appearing at the other end will be
   (*a*) $\frac{1}{6}$ V
   (*b*) $\frac{1}{3}$ V
   (*c*) $\frac{1}{2}$ V
   (*d*) 1 V

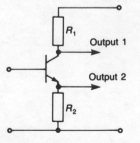

Fig. Q.4

10 Refer to Fig. Q.4. In the basic phase-splitter circuit shown, the ratio of R_1 to R_2 must be
   (*a*) 1 : 2
   (*b*) 1 : 1
   (*c*) 2 : 1
   (*d*) 6 : 1

11 If a resistance is shunted across the tuned load of a radio frequency amplifier, the effect will be to increase the
   (*a*) gain of the amplifier
   (*b*) bandwidth of the amplifier
   (*c*) selectivity of the amplifier
   (*d*) tuned frequency of the amplifier

12 The low frequency response of an amplifier is greatly improved if the coupling is by
   (*a*) resistor and capacitor
   (*b*) transformer
   (*c*) choke and capacitor
   (*d*) direct connection

13 If a typical domestic receiver is tuned to 1100 kHz and has an i.f. of 470 kHz, the frequency of the local oscillator is
   (*a*) 470 kHz
   (*b*) 1100 kHz
   (*c*) 1570 kHz
   (*d*) 10 MHz

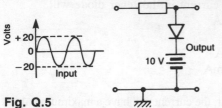

Fig. Q.5

14 Refer to Fig. Q.5. Assuming an ideal diode, the output waveform produced by the circuit shown will be

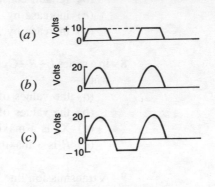

15 Refer to Fig. Q.6. The binary counter has been re-set to zero. After 11 input pulses, the output states will be

Outputs

	4	3	2	1
(*a*)	1	0	1	1
(*b*)	1	1	0	0
(*c*)	1	1	0	1
(*d*)	0	0	0	1

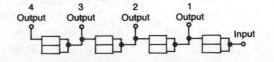

Fig. Q.6

16 A binary counter which uses bistable elements only is required to count 12 pulses. The minimum number of bistables required to do this is
   (*a*) 3
   (*b*) 4
   (*c*) 6
   (*d*) 12

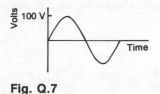

Fig. Q.7

17 If the area of the plates of a capacitor and the distance between them are both doubled the capacitance will be
   (a) reduced by one-third
   (b) unaltered
   (c) tripled
   (d) increased by 6 times

18 Refer to Fig. Q.7. The instantaneous value of the voltage after $\frac{1}{4}$ cycle will be
   (a) 0 V
   (b) +50 V
   (c) +70 V
   (d) +100 V

19 One application of a unijunction transistor is
   (a) voltage amplifier
   (b) differentiator
   (c) pulse generator
   (d) phase shift oscillator

20 If three capacitors of different values are connected in series across a 60 V d.c. supply, the voltage across each capacitor will be
   (a) 20 V
   (b) 60 V
   (c) inversely proportional to its capacitance
   (d) directly proportional to its capacitance

21 The maximum power rating of a transistor may be exceeded
   (a) if the supply voltage is low
   (b) if heat sinks are used
   (c) for short periods of time
   (d) if it is used in a push—pull arrangement

22 Refer to Fig. Q.8. The oscillator used is known as
   (a) Hartley
   (b) Tuned collector
   (c) Tuned base
   (d) Blocking

23 Refer to Fig. Q.8. The output waveform from the circuit shown is
   (a) sinusoidal
   (b) saw-tooth
   (c) square
   (d) pulse

24 When tested, an electrolytic capacitor is found to have no leakage current. This indicates that the capacitor
   (a) is s/c
   (b) is in good condition
   (c) is o/c
   (d) has a low d.c. working voltage

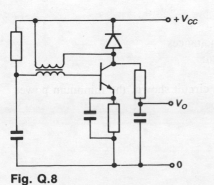

Fig. Q.8

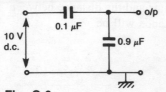

Fig. Q.9

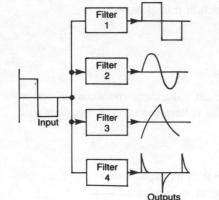

Fig. Q.10

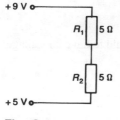

Fig. Q.11

25 Refer to Fig. Q.9. If the input is 10 V d.c., the output voltage is
   (a) 0 V d.c.
   (b) 1 V d.c.
   (c) 9 V d.c.
   (d) 10 V d.c.

26 The impedances presented by a field effect transistor used in the common source configuration are
   (a) low input and low output
   (b) low input and high output
   (c) low input and medium output
   (d) high input and medium output

27 Refer to Fig. Q.10. A square waveform is applied to the inputs for four low pass filters. Which filter has the highest cut-off frequency?
   (a) 1
   (b) 2
   (c) 3
   (d) 4

28 If the readings obtained on a power supply are 100 V on no load and 90 V on full load, the regulation of the power supply is
   (a) 5%
   (b) 9%
   (c) 10%
   (d) 11%

29 Which circuit connection would be used to match a high impedance circuit to a low impedance load?
   (a) A common emitter
   (b) A grounded grid
   (c) An emitter follower
   (d) A common source

30 The low frequency response of a fet audio amplifier is limited by the
   (a) coupling capacitor
   (b) leakage current
   (c) inter-electrode capacitances
   (d) value of load resistor

31 Refer to Fig. Q.11. In the circuit shown, the minimum power rating of resistor R_1 will be
   (a) $\frac{1}{8}$ W
   (b) $\frac{1}{4}$ W
   (c) $\frac{1}{2}$ W
   (d) 1 W

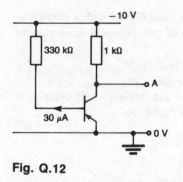

Fig. Q.12

32 Refer to Fig. Q.12. The base voltage is
  (a) −10 V
  (b) −9.9 V
  (c) −5 V
  (d) −0.1 V

33 Refer to Fig. Q.12. The voltage at A is −7 V. The current amplification factor of the transistor is
  (a) 100
  (b) 200
  (c) 300
  (d) 400

34 Refer to Fig. Q.13. The purpose of C_1 is to provide
  (a) r.f. filtering
  (b) interference suppression
  (c) inter-stage coupling
  (d) bias decoupling

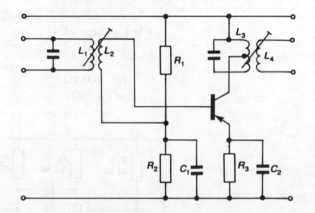

Fig. Q.13

35 Refer to Fig. Q.13. The purpose of the tapping point on L_3 is to
  (a) increase the bandwidth
  (b) improve the selectivity
  (c) tune the amplifier
  (d) provide feedback

36 Refer to Fig. Q.14. The counter shown will divide by
  (a) 6
  (b) 8
  (c) 14
  (d) 16

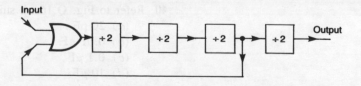

Fig. Q.14

37 An n-channel fet is working with class C bias with a sinusoidal input voltage. During which part of the cycle will drain current flow?

(*a*) The peak of the positive half cycle

(*b*) The whole of the positive half cycle

(*c*) The whole of the positive and negative half cycle

(*b*) The peak of the negative half cycle

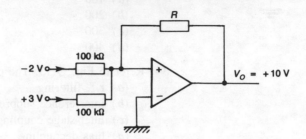

Fig. Q.15

38 In the operational amplifier shown in Fig. Q.15 the resistor *R* has a value of

(*a*) 1 kΩ

(*b*) 10 kΩ

(*c*) 100 kΩ

(*d*) 1 MΩ

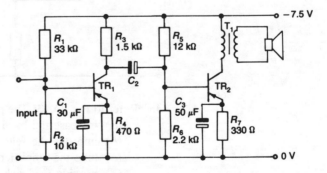

Fig. Q.16

39 Refer to Fig. Q.16. The purpose of C_3 is

(*a*) coupling

(*b*) decoupling

(*c*) i.f. smoothing

(*d*) d.c. biasing

40 Refer to Fig. Q.16. A suitable capacitor C_2 is

(*a*) 20 pF

(*b*) 0.01 μF

(*c*) 0.1 μF

(*d*) 10 μF

41 Refer to Fig. Q.16. If the collector of TR_1 has a voltage of 4.5 V, TR_1 current is

(*a*) 5 mA
(*b*) 2 mA
(*c*) 1 mA
(*d*) 0.5 mA

42 A bistable (flip-flop) has

(*a*) one steady state
(*b*) two steady states
(*c*) three steady states
(*d*) no steady state

43 Which 2-input gate performs the function $F = \overline{A.B}$?

(*a*) AND
(*b*) OR
(*c*) NAND
(*d*) NOR

44 A power loss of 3 dB is represented by a power ratio of

(*a*) $\frac{1}{3}$
(*b*) $\frac{1}{2}$
(*c*) 2
(*d*) 3

45 Which of the following gives truth table of the combination logic circuit shown in Fig. Q.17?

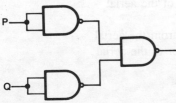

P

Q

Fig. Q.17

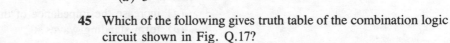

(*a*)

P	Q	F
0	0	1
0	1	0
1	0	0
1	1	0

(*b*)

P	Q	F
0	0	1
0	1	1
1	0	1
1	1	0

(*c*)

P	Q	F
0	0	0
0	1	1
1	0	1
1	1	1

(*d*)

P	Q	F
0	0	0
0	1	0
1	0	0
1	1	1

46 The gate shown in Fig. Q.18 is

(*a*) AND
(*b*) NOR
(*c*) Ex-OR
(*d*) Ex-NOR

Fig. Q.18

47 A fet amplifier in which drain current flows for less than half a cycle of the input signal waveform is termed to be operating in class

(a) A
(b) B
(c) AB
(d) C

48 The Boolean expression for the Ex-OR function is

(a) $(A+\overline{B}).(\overline{A}+B)$
(b) $\overline{A}.B + A.\overline{B}$
(c) $\overline{\overline{A}.B + A.\overline{B}}$
(d) $\overline{A}.\overline{B} + A.B$

49 In a 7-segment display a fault causes the figure eight to be displayed as nine. Which one of the following will also occur?

(a) Three displayed as nine
(b) Nine displayed as three
(c) Six displayed as five
(d) Blank display

50 A receiving aerial should be matched to a transmission cable in order to

(a) increase the impedance of the aerial
(b) form standing waves
(c) draw maximum power from the aerial
(d) transfer maximum power to the aerial

51 Logic analysers are used to test

(a) audio amplifiers
(b) video systems
(c) radio receivers
(d) microprocessor systems

52 Figure Q.19 shows a 7-segment display arrangement. Block Y is a

(a) BCD to 7-segment decoder
(b) driver
(c) counter
(d) latch

53 In a switched mode power supply, which of the following mark-to-space ratios of the switching pulse will produce the lowest voltage at the output?

(a) 1 : 2
(b) 1 : 4
(c) 2 : 1
(d) 4 : 1

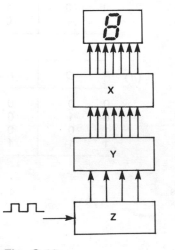

Fig. Q.19

Fig. Q.20

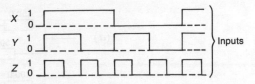

54 Refer to Fig. Q.20. If the inputs, *X, Y* and *Z* to the gate are as shown, the output will be

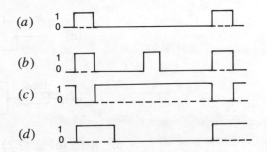

55 Refer to Fig. Q.21. Which combination of binary inputs would produce the output shown?

	A	*B*	*C*	*D*
(*a*)	0	0	0	0
(*b*)	1	0	1	0
(*c*)	0	1	0	0
(*d*)	1	1	0	0

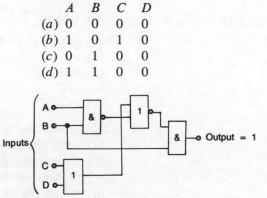

Fig. Q.21

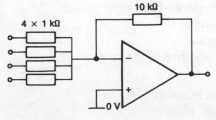

Fig. Q.22

56 Refer to Fig. Q.22. If all inputs are held at 0.2 V, the output voltage will be
 (*a*) +0.8 V
 (*b*) −0.8 V
 (*c*) +8.0 V
 (*d*) −8.0 V

57 In a *S−R* latch, which of the following output combinations is produced when *S* is high and *R* is low?
 (*a*) $Q = 1, \bar{Q} = 1$
 (*b*) $Q = 0, \bar{Q} = 1$
 (*c*) $Q = 1, \bar{Q} = 0$
 (*d*) $Q = 0, \bar{Q} = 0$

58 Which one of the following combinations gives a resultant resistance of 75 Ω?

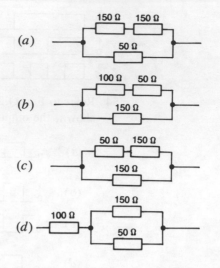

(a)   150 Ω   150 Ω   50 Ω

(b)   100 Ω   50 Ω   150 Ω

(c)   50 Ω   150 Ω   150 Ω

(d)   100 Ω   150 Ω   50 Ω

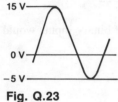

Fig. Q.23

59 Refer to Fig. Q.23. The average voltage of the sine wave shown is
  (a) 0 V
  (b) −5 V
  (c) 5 V
  (d) 10 V

60 The reactance X_c of a capacitor C is given as
  (a) ωC
  (b) $1/\omega C$
  (c) ω/C
  (d) C/ω

61 Applying negative feedback to an amplifier will
  (a) lower the gain
  (b) increase the gain
  (c) produce oscillations
  (d) reduce the bandwidth

62 Forward a.g.c. uses transistors whose gain
  (a) increases with current
  (b) decreases with current
  (c) stays constant irrespective of the current
  (d) is very low

63 High definition television receivers have
  (a) over 1000 lines per picture
  (b) a large screen size
  (c) increased e.h.t. voltage
  (d) more than three primary colours

64 A 20 000 Ω/V voltmeter has a full-scale deflection current of
  (a) 20 μA
  (b) 50 μA
  (c) 20 mA
  (d) 50 mA

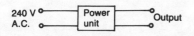

240 V A.C. — Power unit — Output

Fig. Q.24

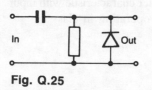

In — Out

Fig. Q.25

65 Refer to Fig. Q.24. The no-load voltage is
  (*a*) 240 V
  (*b*) 350 V
  (*c*) 480 V
  (*d*) 700 V

66 Refer to Fig. Q.25. The output waveform produced by the circuit shown is

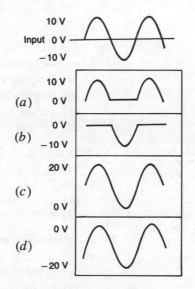

67 Interstage coupling capacitors in a.f. circuits employing unipolar transistors have much smaller values than those using bipolar transistors because bipolar transistors have
  (*a*) low input impedance
  (*b*) high input impedance
  (*c*) high leakage current
  (*d*) low gain

68 Refer to Fig. Q.26. The maximum safe voltage that may be applied across A and B is
  (*a*) 5 V
  (*b*) 10 V
  (*c*) 15 V
  (*d*) 20 V

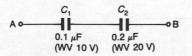

A — C_1 — C_2 — B
0.1 μF (WV 10 V)   0.2 μF (WV 20 V)

Fig. Q.26

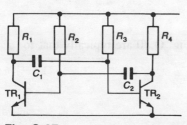

Fig. Q.27

69 Refer to Fig. Q.27. The name given to the oscillator circuit shown is
  (*a*) Colpitts
  (*b*) Hartley
  (*c*) astable multivibrator
  (*d*) bistable multivibrator

70 Refer to Fig. Q.27. If R_2 is half the value of R_3 and C_1 is half the value of C_2, the mark–space ratio of the current will be

(a) 1 : 2
(b) 1 : 1
(c) 2 : 1
(d) 4 : 1

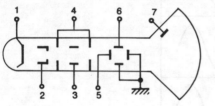

Fig. Q.28

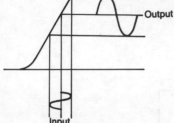

Fig. Q.29

71 Refer to Fig. Q.28. This shows a transfer characteristic with input and output signals. The amplifier is operating in class
(a) A
(b) B
(c) AB
(d) C

72 Lissajous figures are used to measure
(a) frequency
(b) amplitude
(c) period
(d) current

73 Refer to Fig. Q.29. The brightness and focus electrodes are
(a) 1, 2
(b) 2, 3
(c) 2, 6
(d) 1, 7

74 A device which emits electrons when light falls upon it is known as a
(a) photoconductive cell
(b) photovoltaic cell
(c) photo diode
(d) photo-emissive cell

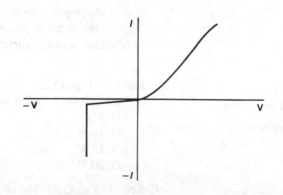

Fig. Q.30

75 The characteristics shown in Fig. Q.30 are typically that for a
(a) SCR
(b) diac
(c) zener
(d) transistor

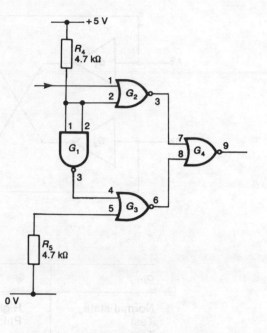

Fig. Q.31

76 Refer to Fig. Q.31. The logic levels at Pin 1 (G_2) and Pin 5 (G_3) are

(*a*) Pin 1, Low; Pin 5, High
(*b*) Pin 1, Low; Pin 5, Low
(*c*) Pin 1, High; Pin 5, Low
(*d*) Pin 1, High; Pin 5, High

77 The truth table of the logic circuit in Fig. Q.32 is

(*a*)	A	B	F
	0	0	0
	0	1	1
	1	0	1
	1	1	1

(*b*)	A	B	F
	0	0	0
	0	1	1
	1	0	1
	1	1	0

(*c*)	A	B	F
	0	0	1
	0	1	1
	1	0	1
	1	1	0

(*d*)	A	B	F
	0	0	1
	0	1	0
	1	0	0
	1	1	1

78 Refer to Fig. Q.32 which shows a part of a digital circuit. A logic probe and pulser were used to test Gate 3. With the pulser placed and activated at each input in turn and the output of the Gate (Pin 6) monitored by the probe, the following results were obtained.

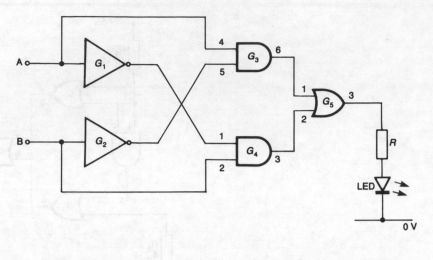

Fig. Q.32

Pin	4	5	6
Normal state	High	High	High
Test	Pulser	—	Observed change
Test	—	Pulser	No change

The faulty pin is
(a) Pin 4: stuck-at-0
(b) Pin 4: stuck-at-1
(c) Pin 5: stuck-at-0
(d) Pin 5: stuck-at-1

79 The advantage of CMOS logic over the TTL type is
(a) TTL consumes less power
(b) CMOS consumes less power
(c) TTL is faster than CMOS
(d) CMOS can accommodate more inputs

80 Logic analysers are used to test
(a) microprocessor systems
(b) digital boards
(c) radio receivers
(d) electric heaters

Appendix I
Resistors

Resistors are the most commonly used components in all electronic circuits.

Stability   Resistors can be divided into two types; high stability and general purpose resistors. A high stability resistor retains its original value throughout its life. The general purpose resistor may change from its specified value during use. Stability is given as a percentage of the nominal or coded value of the resistor.

Tolerance   Tolerance indicates the accuracy of the resistor. As it is expensive to manufacture resistors with very accurate values, they are produced by mass production with certain tolerances. Tolerance is also given as a percentage of the nominal value. For example, a 1 kΩ resistor having a tolerance of $\pm 10\%$ means that its real value could be anywhere between (1 kΩ $-$ 10%) and (1 kΩ $+$ 10%).
Since 10% of 1 kΩ is

$$\frac{10}{100} \times 1000 = 100 \ \Omega$$

then the real value is between

$$1000 - 100 = 900 \ \Omega \quad \text{and} \quad 1000 + 100 = 1100 \ \Omega$$

Power rating   The main limitation of a resistor is its power rating. The power rating of a resistor is the maximum power it can dissipate (as heat) without damaging the resistor. Power rating is given in watts and can be estimated by the physical size of the resistor. Large resistors can dissipate greater power than small ones without overheating.

Example   What is the maximum current that a 100 Ω resistor can take if it has a power rating of $\frac{1}{2}$ W?
Since power $= I^2 R$,

$$0.5 \text{ W} = I^2 \times 100 \ \Omega$$

i.e.

$$I^2 = \frac{0.5}{100} = \frac{5}{1000} = \frac{1}{200}$$

$$I = \sqrt{\frac{1}{200}} = \frac{1}{\sqrt{200}} = \frac{1}{10\sqrt{2}} = \frac{1}{10 \times 1.4} = \frac{1}{14}$$

Therefore $I = 0.07$ A or $0.07 \times 1000 = 70$ mA.

Types of resistors

Type	Power rating	Stability	Tolerance	Other points
Wire-wound	High (up to 300 W)	High (1%)	5%	Large size Expensive
Carbon	Low (up to 3 W)	Low (20%)	5–20%	Cheap Small size Widely used
Metal or oxide film (precision)	Low (less than 2 W)	High (1%)	Very close	Expensive Small size
Metal or oxide film (power)	High (up to 300 W)	High (1%)	Close	Large size Expensive

Resistivity   Different materials present different resistance to the flow of electric current. A conductor such as silver, copper or aluminium has a large number of 'loose' or free electrons allowing almost unrestricted current flow. An insulator such as dry wood, rubber or glass, on the other hand, has very few free electrons and presents a very high resistance to the flow of current. Materials between these two extremes have a corresponding degree of resistance. These differences are expressed by what is known as resistivity — defined as the resistance of a unit cube of the material.

Consider a wire 10 cm long having a resistance of 1 Ω. A length of 20 cm of the same wire will have twice the resistance, i.e. 2 Ω. The resistance of a conductor thus increases with its length. In other words, resistance R is proportional to the length ℓ.

The resistance of a wire is also determined by its cross-sectional area a. In the previous example, suppose the cross-sectional area a = 1 mm $\times$ 1 mm = 1 mm^2. If the cross-sectional area is now doubled to 2 mm^2 it is equivalent to placing two lengths of the original wire side by side. The current now has twice the area to flow through. The resistance that the wire presents to the flow of current is therefore halved.

In general, then, the resistance of a conductor decreases as its cross-sectional area increases. Resistance R is therefore inversely proportional to cross-sectional area a.

For both length and cross-sectional area, $R \propto \ell/a$.

The effect of heat on resistors The resistance of almost every material increases with increasing temperature. The resistances of copper, aluminium, silver and iron for instance increase as they are heated. Such materials are said to have a *positive temperature coefficient* (PTC).

Carbon, which is widely used for ordinary resistors, is the only common material that has a *negative temperature coefficient*. The resistance of carbon thus decreases as its temperature increases.

Resistor colour code The nominal value of a resistor and its tolerance are given by a colour code. A resistor is marked by having up to 4 colour bands around its body, as shown in Fig. A1.1. The resistor value is given by the first 3 bands, as detailed in Table A1.1. Tolerance is given by the fourth band.

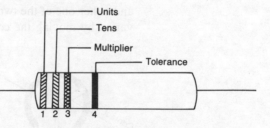

Fig. A1.1 Colour code marking on a resistor

Table A1.1

Colour	Band 1 (value of 1st digit) (units)	Band 2 (value of 2nd digit) (tens)	Band 3 (number of noughts) (multiplier)	Band 4 (tolerance)
Black	0	0	0	
Brown	1	1	1	
Red	2	2	2	
Orange	3	3	3	
Yellow	4	4	4	
Green	5	5	5	
Blue	6	6	6	
Violet	7	7	7	
Grey	8	8	8	
White	9	9	9	
Gold				5%
Silver				10%
None				20%

For resistor values **under 10 Ω**, the colour code of the multiplier (third) band is as follows:

gold 10^{-1} (or 0.1)
silver 10^{-2} (or 0.01)

For example, a 2.2 Ω resistor has the band colours indicated in Fig. A1.2.

Where a fifth band is used, this indicates the temperature coefficient.

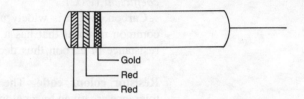

Fig. A1.2   Band colours for 2.2Ω resistor

- Gold
- Red
- Red

Variable resistors   Fig. A1.3 shows diagrammatically a variable resistor (also known as a POT) having three terminals, A, B and C. The resistance between A and B is constant and equal to the total resistance. A variable resistance exists between the centre terminal and either one of the two end terminals, A and B. The resistance is varied by moving the contact arm around the track.

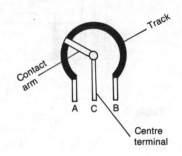

Track

Contact arm

A   C   B

Centre terminal

Fig. A1.3   Variable resistor

Appendix II
Capacitors

Types of capacitors

Type	Capacitance range	Frequency range	Other points
Paper	0.001–1 μF	Medium (50 Hz–1 MHz)	Cheap
Polystyrene	0.0001–1 μF	Medium	Medium stability (1–5%)
Ceramic	up to few μF	v.h.f. and u.h.f.	Small size High stability
Mica and silver mica	a few pF to 0.05 μF	r.f. (200 MHz or over)	Small size Robust High stability Expensive
Air	very low (up to 0.0005 μF)	v.h.f. and u.h.f.	Bulky Widely used as variable tuning capacitors
Electrolytic	high (up to 1000s μF)	Low (d.c. to a.f.)	Cheap Very high tolerance (over 50%) Widely used

Voltage rating The voltage rating of a capacitor is the maximum voltage that the capacitor can be subjected to before the dielectric breaks down. If an a.c. voltage is used, its *peak* value must be below the voltage rating, otherwise the capacitor may be damaged.

Frequency range Apart from the correct voltage rating, a capacitor must have the correct frequency range before it can be successfully used in a circuit. Frequency range is the range within which a capacitor may be used before losses and hence distortion occur.

Electrolytic capacitor Electrolytic capacitors are cheap, small in size, and have large capacitances. They are thus very widely used. They have a definite polarity which must be observed when connecting the capacitor in a circuit.

Appendix III
List of SI units

Angle	radians
Angular velocity	rad/s
Capacitance	farad F
Charge	coulomb C
Current	ampere A
Electromotive force or electric potential	volt V
Energy	joule J
Force	newton N
Frequency	hertz Hz
Impedance	ohm Ω
Inductance	henry H
Length or distance	metre m
Magnetic flux	weber Wb
Magnetic flux density	tesla T
Mass	kilogramme kg
Power	watt W
Pressure	N/m^2
Reactance	ohm Ω
Resistance	ohm Ω
Time	second s
Torque	newton-metre N m
Velocity or speed	m/s
Weight	newton N
Work	joule J

Appendix IV
Components and symbols

RESISTORS

Variable resistor	Preset resistor	Potential divider (potentiometer)

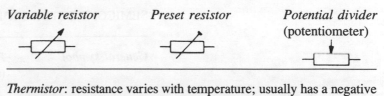

Thermistor: resistance varies with temperature; usually has a negative temperature coefficient; used to compensate change in resistor values due to change in temperature (e.g. car radios).

Negative temperature coefficient resistor (e.g. carbon resistors)	Positive temperature coefficient (PCT) resistor (e.g. wire wound resistor)

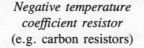

Voltage dependent resistor (VDR): maintains a constant voltage across it as current increases

INDUCTORS

General	Inductor with magnetic core	Variable inductor	Preset inductor

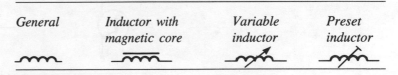

TRANSFORMERS

General symbol (a step-up transformer having a turns ratio of 1:2)	Transformer with magnetic core (a step-down transformer having a turns ratio of 5:1)	Transformer with two secondary outputs (a turns ratio of 1:5 between primary and each secondary)

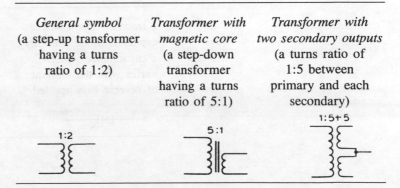

CAPACITORS

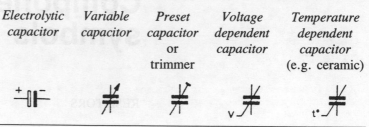

Electrolytic capacitor	Variable capacitor	Preset capacitor or trimmer	Voltage dependent capacitor	Temperature dependent capacitor (e.g. ceramic)

SEMICONDUCTOR DIODES

General symbol	Photo diode (sensitive to light rays)	Zener diode (always used at its reverse break-down voltage which remains constant at all currents; used for voltage stabilisation)

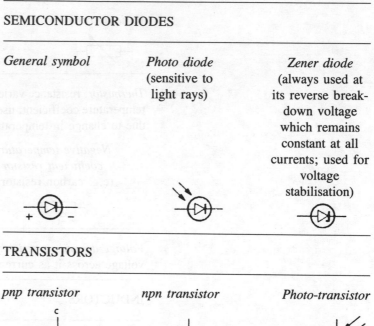

TRANSISTORS

pnp transistor   *npn transistor*   *Photo-transistor*

OTHER DEVICES

Varactor The variable capacitor diode presents a capacitance which varies with the amount of reverse bias applied

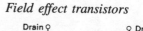

Anode   Cathode

Field effect transistors

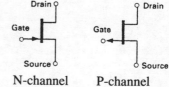

N-channel   P-channel

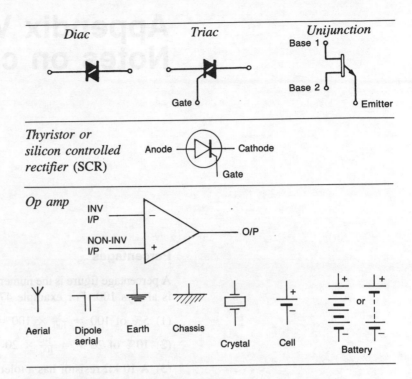

Diac   *Triac*   *Unijunction*

Base 1

Gate

Base 2

Emitter

Thyristor or silicon controlled rectifier (SCR)

Anode   Cathode

Gate

Op amp

INV I/P

NON-INV I/P

O/P

Aerial   Dipole aerial   Earth   Chassis   Crystal   Cell   or   Battery

CONNECTORS AND SWITCHES

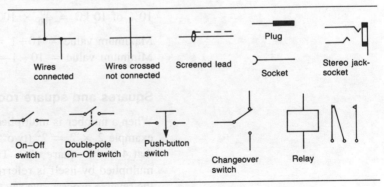

Wires connected   Wires crossed not connected   Screened lead   Plug   Socket   Stereo jack-socket

On–Off switch   Double-pole On–Off switch   Push-button switch   Changeover switch   Relay

COMPONENT ORIENTATION

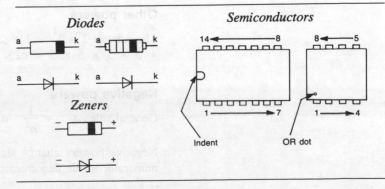

Diodes   *Semiconductors*

a   k   a   k

a   k   a   k

14   8   8   5

1   7   1   4

Indent   OR dot

Zeners

–   +

–   +

Appendix V
Notes on calculations

Percentages

A percentage figure is the numerator of a fraction whose denominator is always 100. For example $47\% = \frac{47}{100}$

(1) 5% of $100 = \frac{5}{100} \times 100 = 5$

(2) 10% of $20\ V = \frac{10}{100} \times 20 = 2\ V$

(3) A $10\ k\Omega$ resistor has a tolerance of $\pm 10\%$. State the maximum and the minimum values that it can have.

10% of $10\ k\Omega = \frac{10}{100} \times 100 = 1\ k\Omega$

Maximum value $= 10+1 = 11\ k\Omega$
Minimum value $= 10-1 = 9\ k\Omega$

Squares and square roots

When a number is multiplied by itself, it is said to be squared. For example $2 \times 2 = 2^2$ (two 'squared') which is equal to 4. We say that 4 is the square of 2. The number of times that the number is multiplied by itself is referred to as the power — in this case 2. In the reverse process, we say that 2 is the square root of 4 or $2 = \sqrt{4}$.

Other powers

The cube of 2 is $2^3 = 2 \times 2 \times 2 = 8$. 5 to the fourth power is $5^4 = 5 \times 5 \times 5 \times 5 = 625$.

Negative powers

General rule: $a^{-2} = \frac{1}{a^2}, \quad a^{-3} = \frac{1}{a^3},$ etc. $3^{-2} = \frac{1}{3^2} = \frac{1}{9}$

Negative powers change sign (to positive) when moved from the numerator to the denominator, and vice versa.

$$2^{-3} = \frac{1}{2^3} = \frac{1}{8} \qquad \frac{1}{3^{-2}} = 3^2 = 9$$

Any number lifted to the power of zero is equal to one: $2^0 = 1$, $10^0 = 1$.

Powers of ten

An easy way of dealing with calculations involving many decimal points is the use of the powers of ten.

$$10^0 = \qquad 1 \qquad 10^{-1} = \frac{1}{10} = 0.1$$

$$10^1 = \qquad 10 \qquad 10^{-2} = \frac{1}{100} = 0.01$$

$$10^2 = \qquad 100 \qquad 10^{-3} = \frac{1}{1000} = 0.001$$

$$10^3 = \qquad 1\ 000 \qquad 10^{-4} = \frac{1}{10\ 000} = 0.0001$$

$$10^4 = \qquad 10\ 000 \qquad 10^{-5} = \frac{1}{100\ 000} = 0.000\ 01$$

$$10^5 = \qquad 100\ 000 \qquad 10^{-6} = \frac{1}{1\ 000\ 000} = 0.000\ 001$$

$$10^6 = 1\ 000\ 000$$

For example, $2000\ V = 2 \times 1000\ V = 2 \times 10^3\ V$ and $0.01\ A = 10^{-2}\ A$.

In multiplying, powers of ten are added:

$$10^2 \times 10^1 = 10^{2+1} = 10^3 \qquad 10^4 \times 10^{-1} = 10^{4-1} = 10^3$$

Multiples and sub-multiples of units

Multiplication factor		Prefix	Symbol
1 000 000 000	10^9	Giga	G
1 000 000	10^6	Mega	M
1 000	10^3	kilo	k
0.001	10^{-3}	milli	m
0.000 001	10^{-6}	micro	μ
0.000 000 001	10^{-9}	nano	n
0.000 000 000 001	10^{-12}	pico	p

EXAMPLE 1

If the current through a 4.7 kΩ resistor is 2 mA, what would be the voltage drop across the resistor?

$$4.7 \text{ k}\Omega = 4.7 \times 10^3 \ \Omega \text{ and } 2 \text{ mA} = 2 \times 10^{-3} \text{ A}$$
$$V = IR = 2 \times 10^{-3} \times 4.7 \times 10^3 = 2 \times 4.7 \times 10^{-3} \times 10^3$$
$$= 2 \times 4.7 \times 10^0 = 2 \times 4.7 \times 1 = 2 \times 4.7 = 9.4 \text{ V}$$

Appendix VI
The decibel

The decibel (dB) is a logarithmic unit for expressing the ratio of two powers, such as power gain. The decibel is a more suitable unit to use than the numerical gain ratio because it provides a smaller number for the gain. It is also more convenient to use in a.f. applications because the human ear responds to sound intensities in a logarithmic manner.

For the amplifier shown in Fig. A6.1 where p_{in} and p_o are the input and output powers, the power gain G_p in dB is given as

$$G_p = 10 \log_{10} \frac{\text{Output power}}{\text{Input power}} = 10 \log_{10} \frac{p_o}{p_{in}} \text{ dB}$$

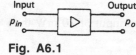

Fig. A6.1

Since p_o/p_{in} = power gain, then
gain in dB = $10 \log_{10}$ (power gain).

It should be noted that the decibel is a unit of power ratio and not a measure of the power level.

The gain in dB may be either positive or negative. A positive dB value represents a gain ratio greater than 1 (p_o greater than p_{in}), while a negative dB value represents a gain ratio smaller than 1 (p_o smaller than p_{in} i.e. a *loss*).

To calculate the gain in dB using the formula given above requires a knowledge of logarithms. It is therefore useful to memorise the basic conversions given in Table A6.1 which, as we shall see later, may be used to calculate most of the conversions that are likely to crop up.

Table A6.1

dB gain	Power gain
0	1
3	2
10	10
20	100 (10^2)
30	1000 (10^3)

Table A6.2

dB gain	Voltage or* current gain
0	1
3	$\sqrt{2} = 1.4$
6	2
20	10
40	100
60	1000

* Assumes equal resistances

It should be noted that (*a*) for a gain of 1 the dB gain is zero and (*b*) that a 10 times increase in gain (e.g. from 100 to 1000) the dB gain increases by 10 dB (from 20 to 30). For a loss, the dB value is negative. For example, a gain of $\frac{1}{2} = -3$ dB and $1/10 = -10$ dB and so on.

EXAMPLES

(1) If an amplifier has a gain of 3 dB, and an input of 10 mW then the output may be calculated as follows:

3 dB = a power gain of 2

Hence, Output = 2 × input = 2 × 10 mW = 20 mW

(2) Refer to Fig. A6.2. To find the input power:

10 dB = a power gain of 10

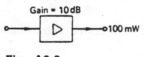

Fig. A6.2

Hence, Input power $= \dfrac{\text{Output power}}{10} = \dfrac{100 \text{ mW}}{10} = 10 \text{ mW}$

Voltage and current gains

Both voltage and current gains of a circuit can be calculated in decibels, *provided* the input and load resistances are of *equal* value, as follows:

$$\text{Gain in dB} = 20 \log \frac{v_o}{v_{in}} = 20 \log \frac{i_o}{i_{in}}$$

where v_o and i_o are the output, and v_{in} *and* i_{in} are the input voltage and current respectively.

From Table A6.2 -3 dB = a voltage gain of $1/1.4 = 0.7$ or 70%. For example, if a 3 dB attenuator (i.e. having a dB loss of 3 dB) has an input voltage of 10 mV then the output is $0.7 \times 10 = 7$ mV.

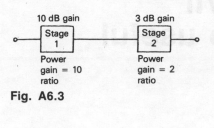

Fig. A6.3

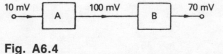

Fig. A6.4

Multi-stage gain

One further advantage of using decibels is that the gain in dB of an amplifier having more than one stage is equal to the sum of the dB gain of each stage. Fig. A6.3 shows a two-stage amplifier. Stage 1 has a gain of 10 dB (i.e. a power gain of 10) and stage 2 has a gain of 3 dB (or a power gain of 2). The overall gain is 10 dB + 3 dB = 13 dB. The overall power gain ratio on the other hand is given as the product of the gain of each stage, namely $10 \times 2 = 20$.

EXAMPLE Refer to Fig. A6.4

Stage A gain = 100/10 = 10, or 20 dB (Assuming $R_{in} = R_L$)
Stage B gain = 70/100 = 0.7, or −3 dB
Overall gain = 70/10 = 7. Overall gain in dB = 20−3
$$= 17 \text{ dB}$$

Conversion from decibels to ratios or vice versa for values other than those given in the two tables may be worked out by sub-division into suitable stages. For example, 13 dB may be divided into 10 dB + 3 dB, and 10 dB is a ratio of 10 and 3 dB is a ratio of 2. Hence the total equivalent ratio is $10 \times 2 = 20$ (refer to Fig. A6.3).

EXAMPLES

(3) If the input to a device is 10 mW and its output is 1 W what is the gain in dB?

Solution

$$\text{Power gain} = \frac{1 \text{ W}}{10 \text{ mW}} = \frac{1000 \text{ mW}}{10 \text{ mW}} = 100$$

A power gain of 100 = 20 dB.

(4) Refer to Fig. A6.5. Find (*a*) the overall gain, (*b*) the overall gain in dB, (*c*) the gain of stage B in dB and as a ratio.

Fig. A6.5

Solution

(*a*)   Overall gain $= \dfrac{v_o}{v_{in}} = \dfrac{100 \text{ mV}}{50 \text{ mV}} = 2$

(*b*)   Voltage gain of 2 = 6 dB.
(*c*)   Gain of stage B = 6 dB − 9 dB = −3 dB.
         Stage B gain as a ratio is 1/1.4 = 0.7.

Appendix VII
Answers to useful exercises

Useful exercises I

1	(b)	21	(c)	41	(a)	61	(b)
2	(d)	22	(d)	42	(a)	62	(b)
3	(b)	23	(a)	43	(c)	63	(c)
4	(a)	24	(c)	44	(b)	64	(d)
5	(b)	25	(c)	45	(b)	65	(b)
6	(c)	26	(b)	46	(c)	66	(a)
7	(a)	27	(c)	47	(c)	67	(c)
8	(d)	28	(a)	48	(b)	68	(c)
9	(c)	29	(d)	49	(d)	69	(a)
10	(b)	30	(b)	50	(c)	70	(d)
11	(d)	31	(d)	51	(d)	71	(a)
12	(c)	32	(b)	52	(a)	72	(b)
13	(b)	33	(a)	53	(b)	73	(b)
14	(a)	34	(b)	54	(c)	74	(b)
15	(b)	35	(b)	55	(c)	75	(c)
16	(c)	36	(c)	56	(b)	76	(d)
17	(b)	37	(c)	57	(a)	77	(b)
18	(b)	38	(b)	58	(c)	78	(a)
19	(a)	39	(a)	59	(a)	79	(a)
20	(d)	40	(c)	60	(a)	80	(b)

Useful exercises II

1	(b)	7	(c)	13	(c)	19	(c)
2	(c)	8	(b)	14	(d)	20	(c)
3	(c)	9	(c)	15	(a)	21	(b)
4	(c)	10	(b)	16	(b)	22	(d)
5	(d)	11	(b)	17	(b)	23	(b)
6	(b)	12	(d)	18	(d)	24	(c)

25	(b)	39	(b)	53	(b)	67	(a)
26	(d)	40	(d)	54	(c)	68	(c)
27	(a)	41	(b)	55	(d)	69	(c)
28	(d)	42	(b)	56	(d)	70	(b)
29	(c)	43	(c)	57	(c)	71	(a)
30	(a)	44	(b)	58	(b)	72	(a)
31	(d)	45	(c)	59	(c)	73	(b)
32	(d)	46	(d)	60	(b)	74	(d)
33	(a)	47	(d)	61	(a)	75	(c)
34	(d)	48	(b)	62	(b)	76	(c)
35	(b)	49	(c)	63	(a)	77	(b)
36	(c)	50	(c)	64	(b)	78	(d)
37	(a)	51	(d)	65	(b)	79	(b)
38	(d)	52	(a)	66	(c)	80	(a)

Index

3-dB point 33
555 timer 209–10
741 op amp 202–3
748 op amp 203

acceptor 39, 121
address 67, 231
 bus 232
aerial 47, 49
alloys 47
alternating current (a.c.) 9
alternator 44
ammeter 233
ampere 1
amplification 31, 136–7
 graphical analysis 138–41
amplifier 31
 R–C coupled 126
 bandwidth 186
 biasing 185
 buffer 37
 cascade 186
 chrominance 101
 common emitter 128–41
 d.c. 192–3
 difference 69, 202
 frequency response 32–7
 gain 31
 i.f. 187
 integrating 200–201
 inverting 198
 luminance 101
 overdriving 52, 139
 power 34, 188
 push-pull 34–5, 188–92
 radio frequency 35–6
 selective 35
 summing 198–9
 testing 240–42
 transformer coupled 139–41
 tuned 35–6
 voltage 34
 wide-band 35
amplitude 9
 distortion 33
amplitude modulation 77, 86
 bandwidth 79
 detector 81–3
analogue instruments 235
analogue-to-digital converter (ADC)
 26, 65, 103
analyser 237, 243
AND 59
angular velocity 25
anode 91, 155
arithmetic and logic unit (ALU) 66
astable 204, 208–9, 214
atom 120
attenuator 38
audio frequency 16
 bandwidth 33
 range 16
 signal generator 37–8
automatic frequency control (a.f.c.)
 88, 158
automatic gain control (a.g.c.)
 87–8
automatic test equipment (ATE)
 238
auto-transformer 46
average 10

back plate 227
band pass filter 54
bandwidth 32, 79, 186
 a.f. 33
 operational amplifier 197, 203
 resonance 40
 telegraph 74
 telephone 73
trasnmission 72
base current 132
bass 195
BBC f.m. transmission 87
BCD to seven-segment 226–7
beam 91, 92, 96
bias 122
 base current 132
 chain 128
 class C 167
 tape recorder 69
 transistor 124
biasing 185
bidirectional 66, 68, 158
binary 63–4, 67
 code 67
 counter 251, 220–22
binary coded decimal (BCD) 227
bipolar 148
bistable 204, 205–7
 fault finding 250–51
 S–R 223
blocking capacitor 134
blocking oscillator 214
boolean expression 218–19
breakdown 154
breakdown voltage 177
bridge rectifier 172
brightness 92
buffer 37, 68
bus 67, 232

camera 96
capacitance 20
capacitive reactance 29
capacitor 19
 blocking 134
 charging 23, 19–20
 coupling 134, 186
 decoupling 135, 136, 186

discharging 23
 energy 20
 interelectrode 186
 parallel 20—22
 reactance 29
 reservoir 173
 resistance 25
 series 20—22
 speed-up 206
 stray 196
 testing 253, 256
 total 21—2
 types 277
carbon 120
carrier 74
cascade 186
cascode 195
cathode 91, 155
cathode ray oscilloscope (CRO)
 93—5, 235
cathode ray tube (CRT) 91, 92,
 98, 102
cell 146
central processing unit (CPU) 230,
 231
centre tap 46, 171
 transformer 189
charge 20
charging 19—20, 23
chip 159
 7400 224
 7402 224
 7404 224
 memory 231
chrominance 101
 processing 103
clamper 52, 185, 167—70
clamping 52
class A 184, 188
class B 184
class C 185
clipper 52, 163
clipping 52, 163—7
clock 232
 digital 65
clocked J—K bistable 223
coding 90
coil 91—2
colour 100
 code 275
 transmission 101
 triad 103
colour television 100—101
 compatibility 101
 receiver 101—2
Colpitt's oscillator 211—12
combinational logic 219

common base 125, 143, 145
common collector 125, 144, 145
common emitter 125, 128, 145
common source 150
comparator 197
compatibility 101
complementary metal oxide silicon
 (CMOS) 217, 227
 gate 218
complementary push-pull 190—92
computer 66
conductor 121
control bus 232
converter 58, 182, 183
coulomb 20
counter 64, 220—23, 227
 feedback 221
 ring 223
coupling 50
 R—C 136, 140—41
 capacitor 134, 186
 opto 147
 transformer 139—41, 143, 187
crystal lattice 121
crystal oscillator 212—13
current 1, 9
 feedback 194
 gain 31
 leakage 130
 measurement 233
 tracer 236
cut off 132—3, 184—5, 240—42
 frequency 186
cycle 13

damped oscillation 41
damping 40, 188
data bus 232
decade counter 223, 227
decibel 285
decimal counter 22
decoder 226
decoder/driver 227
decoupling capacitor 135, 136, 186
deflection 91—2
demodulation 74
denary numbers 63—4
depletion fet 152
depletion layer 122
deviation 84—5
diac 158
dielectric 19
difference amplifier 69, 202
differential amplifier 69
differentiator 54, 162, 164, 197
diffusion 122
digital audio tape-recorder (DAT) 65

digital clock 65
digital display 225—9
digital instruments 235
digital processing 65
digital testing 240, 242—3
digital TV receiver 103—4
digital voltmeter (DVM) 235
digital-to-analogue converter (DAC)
 65, 103
digitised signals 103
diode 122—3
 bidirectional 158
 clipping 163—5
 light emitting 146—7
 photo 146
 protection 214
 steering 206
 switching 154
 testing 253—4, 256—7
 variable capacitance 158
 zener 154
direct current (d.c.) 1
 amplifier 192—3
 restorer 52, 167
discharging capacitor 23
discriminator 88
dissipation 141
distortion 33
divide by two (÷ 2) device 207
divider 64—5
donor 121
double sideband transmission 80
drain 148
drift 37, 193
driver 226
dropper resistor 177
dual-in-line (DIL) 231
dynamic test 238—40
dynamo 44

efficiency, transformer 45
electric field 19—20
electrode 92, 94
electrolytic capacitor 277
 testing 256
electrolytic plating 228
electromagnet 43
electromagnetic deflection 91—2
electromagnetic spectrum 100
electromagnetic waves 75
electromagnetism 42
electromotive force (e.m.f.) induced
 44
electron 1, 120
 beam 91, 92
 gun 91, 92
 lens 91, 92

electronic voltmeter (EVM) 235
electrostatic deflection 91–2
electrostatic field 91
emitter follower 178, 179
energy 20
enhancement fet 152
envelope 77
erase head 69, 70
Exclusive-NOR 61
Exclusive-OR 61, 228
extra high tension (e.h.t.) 76, 100

fading 87
farad 20
fault finding 238–57
  bistable 250–51
  d.c. amplifier 244–5
  digital circuit 251–3
  fet amplifier 245–7
  output stage 248–50
  power supply 247–8
feedback 36–7, 193–4
  binary counter 221
ferrites 47
field 97
  electromagnetic 42
  hold 215
  sync 97
field effect transistor (fet) 148–53
  gate 218
filter 53–4, 61–2
flicker 97
flip-flop 220
flyback 92, 96
follower 199
forward bias 122
forward voltage drop 122, 126
frame 97
free electron 121
frequency 13–14
  amplifier 32–7
  changing 82
  compensation 203
  counter 64
  cut off 186–7
  deviation 84
  discriminator 88
  distortion 33
  drift 37
  intermediate 82
  measurement 236
  meter 64
  pulling 37
  resonance 39
  synthesis 68
frequency modulation (f.m.) 84, 86
  advantages 86

bandwidth 85–6
  side frequency 85–6
front plate 227
full-wave rectifier 53, 56, 171–2
functional testing 238
fundamental 54

gain 31, 286–7
  a.c. 202
  overall 32
  two stage 31–2
gates 59–64, 148, 155, 218
  boolean expression 218–19
  combination 62–3
  symbols 62
  testing 242
  truth tables 62
generator 44
germanium 122, 126
graphical analysis 138–41
gun 91, 92

half-split 239
half-wave rectifier 53, 56, 171
harmonics 17, 54, 161
  complex waves 17–18
  sawtooth 18
  square wave 18
Hartley oscillator 211
head 69, 70
height control 216
henry 23
hertz 13, 15
hi-fi 104, 195
high definition television (HDTV) 104
high frequency 75
high frequency bias 69
high pass filter 53, 161
hole 121, 125
hydrogen 120

impedance 30
  internal 49
  matching 49–51
  ratio 46
impurity 121
in-circuit measurement 238
in-circuit testing 255–7
in-line tube 103
inductance 23
inductive reactance 29
inductor 23
  reactance 29
  resistance 25
  testing 253, 256
input characteristics 125–6

input impedance 49
input/output interface 232
instruction 66
insulator 121
integrated circuit (IC) 159–60
integrating amplifier 200–201
integrator 54, 163, 164, 197
interelectrode capacitor 186
interface 67, 232
interlacing 96–7
intermediate frequency (i.f.) 82
  amplifier 83, 187
  TV receiver 98
internal impedance 49
internal resistance 175, 233
  voltmeter 234
inverter 58, 181
inverting amplifier 198
inverting input 197
isolator 147

J–K bistable 223
joule 20
junction diode 122

keyboard 232
Kirchhoff's laws 6–7

lagging 26
large scale integration (LSI) 159, 217
latch 223
leading 26
leakage current 122, 130
lens 91, 92
light emitting diode (LED) 146–7, 225–7
limiter 52, 163
line 96
  frequency 97
  sync 97
liquid crystal display (LCD) 225, 227–9
Lissajou figures 236
load line 137, 151
loading effect 233–4
logarithmic scale 32
logic analyser 237, 243
logic combination 219
logic level 217
logic probe 236
logic pulser 236
logic test instruments 236–7
long wave 75
loose electron 121
loudspeaker 47, 71
low frequency (LF) 75

low-pass filter 54, 161, 175
lower sideband 79
luminance 101

MOS 148, 160
MOSFET 152
magnetic alloys 47
magnetic circuit 47
magnetic core 43
magnetic field 42–8
magnetic tape 69
mark 15
mark-to-space 15, 183
matching 49
mean 10
measurement 233
medium frequency (MF) 75
medium scale integration (MSI)
    217
medium wave 75
memory 66, 230–31
metal-oxide-silicon (MOS) 227
micro-controller 232
microcomputer 67, 217
    architecture 231–2
microphone 73
microprocessor 60, 217, 231
    speed 232
    testing 237, 243
minicontroller 232
mixer 82
modulation 74
    depth 77
    pulse 89
monochrome 101
monostable 204, 207–8
motor 44
moving coil 47, 235
multimeter 235
multiplexing 90
multivibrator 204–9

NAND 60, 217, 218
negative feedback 36
negative temperature coefficient
    275
neutralising 195
neutron 1
NICAM 103, 104
no-load voltage 173
node 6
noise 86, 90
non-inverting input 197
non-sinusoidal oscillator 214
non-volatile 231
non-zero output 199, 200
NOR 60, 218

NOT 61
n-type semiconductor 122
nucleus 120

offset 199–200
    null 200, 202
ohm 1
Ohm's law 2
ohmmeter 253
operational amplifier 197–203
opto-coupler 147
oscillator 37, 211–16
    local 82
    timebase 93
output characteristics 125, 126–7,
    149
output impedance 49
output stage 37, 248–50
over-modulation 78
overdriving 139
overloading 52
    protection 181
overshoot 214

page 105
parallel connector 67
parallel resonance 39
peak 9
peak inverse voltage 174
peak-to-peak 9
period 14, 15
periodic time 14, 15
    measurement 235
peripheral 67, 232
permeability 47
persistence of vision 96
phase difference 26
    measurement 235
phase discriminator 68
phase shift 157
    oscillator 212
phase splitter 24–35, 189
phase-locked loop 68
phasor diagram 27–30
phasor representation 25–30
photoconductive cell 146
photodiode 146
photoelectric devices 146–7
photo-emissive cell 146
photo-sensitive 147
phototransistor 146
photovoltaic cell 146
pinch-off 149–50
pitch 17
playback head 69
positive feedback 36

positive temperature coefficient (P)
    275
potential barrier 122
potential difference 1, 5
potential divider 3–4
power 7
    a.c. 9
    amplifier 34, 188
    d.c. 9
    dissipation 141
    gain 31
    rating 273–4
power supply 56
    regulated 57–8
pre-amplifier 69
precision-in-line tube 103
primary colour 100
primary winding 44
probe 236
program 66
protection 181
proton 1, 120
p-type semiconductor 122
pulling 37
pulse amplitude modulation 89
pulse code modulation (PCM)
    89–90
pulse position modulation 89
pulse waveform 14
pulse width modulation 89
pulser 236
push-pull 34–51, 84
    amplifier 188–92

Q-factor 40
Q-point 185
quartz crystal 213
quiescent point 127, 138–41

radio bandwidth 83
radio frequency (r.f.)
    amplifier 35–6, 194–5
    tuner 82, 84
radio receiver 81–3, 87–8, 187
    testing 239–40
radio superhet 82
radio transmission 75
radio transmitter 80–82
ramp 200
random access memory (RAM) 231
rating power 273–4
rating voltage 277
R–C oscillator 212
reactance 28–30
READ 230–32
read only memory (ROM) 230–31
receiver, a.m. 82–3

receiver, f.m. 87—8
receiver, television 98—100, 103—4, 187
rectangular wave 14
rectification 52—3, 171—3
rectifier 53, 235
  bridge 172
  diode 171—2
register 222
regulated power supply 176—81
regulation 57, 175—6
regulator 57
rejector 39
relaxation oscillator 214
relay 47 73
reluctance 47
repeater 73
reservoir capacitor 173
resistance 1
  internal 233, 234
  total 3
resistivity 274
resistor dropper 177
resistor testing 253, 255—6
resistor types 274
resistor, parallel 5—6
resistor, series 2
resonance 39—41
  bandwidth 40
reverse bias 185
ring counter 223
ripple 57, 173, 175, 183
root mean square (r.m.s.) 9—12

sampling 90
saturation 132—3, 241, 242
sawtooth 14, 163
  generator 215
scanning 96—7
Schmitt trigger 167
screening 47
secondary winding 44
selective amplifier 35
selectivity 40
semiconductor 121
sensitivity, voltmeter 234
series regulation 178—81
series resonance 39
seven-segment display 225—7
shadow mask tube 102
shielding 47
shift register 222—3
short wave 75
shunt regulator 176—7
side frequency 78
sideband 78—9
  vestigial 98

signal biasing 185
signal generator 36—7
signature analyser 237, 243
silicon 122, 126
silicon chip 159
silicon controlled rectifier (SCR) 155—7, 183
  testing 254—5
sine wave 9, 14
single sideband transmission 80
single stage amplifier testing 240—42
small scale integration (SSI) 217
smoothing 56, 174—5
  electronic 180
soft iron 43, 47
solenoid 43
sound 16—18
  quality 17
source 148
space 15
spectrum, electromagnetic 75
speed control 69
speed-up capacitor 206
spike 54, 162
splitter 51
square wave 14, 18, 54
S—R bistable 223
stabilisation 130
stability 36, 273
static current gain (h_{fe}) 127
static test 238, 240—42
steering diode 206
step-down 45, 50
stepper motor 232
step-up 45
stereophonic sound 104
stray capacitor 186
stuck-at fault 242
stuck-at-one 242
sub-routine 67
substrate 152
summing amplifier 197, 198—9
super high frequency (SHF) 76
super large scale integration (SLSI) 217
sweep 92
switched mode power supply (SMPS) 183
switching diode 154
switching element 183
synchronisation 93
synchronisation pulse 97
synchronising separator 100

tape recorder 69—70
tapped primary 187

tapping 46
telegraph 73—4
telephone 73
teletext 103, 105
television 96—105
  625 system 97
  colour 100—101
  electrodes 94
  i.f. 98
  timebase 100
television receiver 98—100, 187
  digital 103—4
temperature coefficient 275
tesla 47
testing, component 253—7
testing, digital 242—3
testing, dynamic 238—40
testing, functional 238
testing, static 238, 240—42
thermal runaway 130, 148
thermionic valve 91
thermistor 193
thick film 159
thin film 159
thyristor 155—7, 183
time constant 23—5, 162, 163
  CR 23
  LR 24
time delay 208
time multiplexing 90
timebase 92, 93
timer 209—10
toggle 223
tolerance 273
tone 16
  control 195—6
tracer 236
transducer 67, 69, 71
transfer characteristics 125, 127
transformer 44—8, 57
  auto 56
  centre tap 46, 189
  coupling 139—41, 143, 187
  efficiency 45
  impedance ratio 46
  isolation 147
  matching 50
  screening 47
  shield 47
  step-down 45
  step-up 45
  tapped 171
  tapping 46
  testing 253
transistor 124—7
  bipolar 148
  clipping 166

driver 226
faults 241−2
field effect 148−53
load resistor 129
phase splitter 189−90
photo 146
stabiliser 178
switch 133
testing 254, 256−7
unijunction 158
unipolar 148
transistor-transistor logic (TTL)
  217
transmission bandwidth 72
transmission, double sideband 80
transmission, single sideband 80
transmitter, a.m. 81
transmitter, f.m. 86
treble 195
triac 158
triad 103
triangular wave 14
Trinitron 103
truth table 59−62, 223
tuned amplifier 35−6
tuned base oscillator 211
tuned collector oscillator 211
tuner 82, 83

tuning 158

ultra high frequency (UHF) 76
  oscillator 213
unidirectional 147
unijunction 158
  oscillator 215
unipolar 148
unit-under-test 238
upper sideband 79

valve 91
varactor 157−8, 213
variable resistor 276
very high frequency (VHF) 75, 87
very large scale integration (VLSI)
  217
very low frequency (VLF) 75
vestigial sidebands 98
video processing 103
video signal 52
video waveform 97
virtual earth 198
visual display unit (VDU) 67
volatile 231
voltage 1
  amplifier 34
  feedback 194

follower 199
gain 31
measurement 233
voltage controlled oscillator (VCO)
  68
voltmeter 233
digital 235
electronic 235

watt 7
waveform 13
  d.c. 13
  period 114
  types 14
wide-band amplifier 35
working voltage 22
WRITE 230−31, 232

x-plate 92, 93
x-shift 92

y-amplifier 93
y-attenuator 93
y-plate 92, 93
y-shift 92

zener 154
  clipping 166
  regulator 177